Springer Series in Design and Innovation

Volume 65

Springer Series in Design and Innovation (SSDI) publishes books on innovation and the latest developments in the fields of Product Design, Interior Design and Communication Design, with particular emphasis on technological and formal innovation, and on the application of digital technologies and new materials. The series explores all aspects of design, e.g. Human-Centered Design/User Experience, Service Design, and Design Thinking, which provide transversal and innovative approaches oriented on the involvement of people throughout the design development process. In addition, it covers emerging areas of research that may represent essential opportunities for economic and social development.

In fields ranging from the humanities to engineering and architecture, design is increasingly being recognized as a key means of bringing ideas to the market by transforming them into user-friendly and appealing products or services. Moreover, it provides a variety of methodologies, tools and techniques that can be used at different stages of the innovation process to enhance the value of new products and services.

The series' scope includes monographs, professional books, advanced textbooks, selected contributions from specialized conferences and workshops, and outstanding Ph.D. theses.

The volumes of the series are single-blind peer-reviewed.

Keywords: Product and System Innovation; Product design; Interior design; Communication Design; Human-Centered Design/User Experience; Service Design; Design Thinking; Digital Innovation; Innovation of Materials.

How to submit proposals

Proposals must include: title, keywords, presentation (max 10,000 characters), table of contents, chapter abstracts, editors'/authors' CV.

In case of proceedings, chairmen/editors are requested to submit the link to conference website (incl. relevant information such as committee members, topics, key dates, keynote speakers, information about the reviewing process, etc.), and approx. number of papers.

Proposals must be sent to publishing editor Mr. Pierpaolo Riva (pierpaolo.riva@springer.com).

Daniel Raposo · Luis Farinha · João J. Ferreira · Marcelo Calvete Gaspar · Maria Leopoldina Alves · Luis Serrano
Editors

Strategic Design for Sustainable Futures

Approaches in Communication, Fashion and Product Design

Springer

Editors
Daniel Raposo
CIAUD, Research Centre for Architecture, Urbanism and Design, Lisbon School of Architecture, Universidade de Lisboa / CIAUD.IPCB and Polytechnic University of Castelo Branco
Castelo Branco, Portugal

João J. Ferreira
University of Beira Interior
Covilhã, Portugal

Maria Leopoldina Alves
University of Leiria and Oeste
Leiria, Portugal

Luis Farinha
Polytechnic University of Castelo Branco
Castelo Branco, Portugal

Marcelo Calvete Gaspar
University of Leiria and Oeste
Leiria, Portugal

Luis Serrano
University of Leiria and Oeste
Leiria, Portugal

ISSN 2661-8184 ISSN 2661-8192 (electronic)
Springer Series in Design and Innovation
ISBN 978-3-032-33281-3 ISBN 978-3-032-33282-0 (eBook)
https://doi.org/10.1007/978-3-032-33282-0

This Springer imprint is published by the registered company Springer Nature Switzerland AG
The registered company address is: Gewerbestrasse 11, 6330 Cham, Switzerland

Preface

In the context of contemporary design challenges, sustainable innovation has emerged as a foundational condition for socio-economic transformation rather than a peripheral objective. The current era is characterised by complex and interdependent societal challenges, shaped by significant transitions in ecology, digitalisation, demography, and organisational models. These transformations span environmental, economic, social, and institutional domains, necessitating integrative approaches that combine scientific knowledge, technological development, and social responsibility. The Triple Helix framework, which unites academia, industry, and government, provides an effective operational model for locally grounded innovation ecosystems. Its significance lies not in symbolic alignment, but in its ability to connect knowledge production, economic activity, and public policy within concrete innovation processes aimed at long-term value creation.

Design research occupies a distinctive role within these ecosystems. Its multidisciplinary character enables operation in complex scenarios where technical, social, environmental, and cultural dimensions intersect. Through project-based inquiry, design extends beyond the production of artefacts or services to generate situated knowledge rooted in intervention, experimentation, and critical documentation. Within innovation systems, design functions as both an investigative and projective discipline. It frames problems, supports diagnosis, and mediates among stakeholders with divergent interests, identifying explicit and latent needs prior to structuring responses. Strategic considerations are embedded in the definition of the object of study, analytical processes, construction of future scenarios, and materialisation of solutions under conditions of uncertainty.

The digital transition exemplifies this dynamic. It is widely recognised how digitalisation is reshaping business models, communication systems, and modes of participation, thereby renewing collaboration practices and co-creation processes. This book's chapters focus on generative design, digital visual merchandising, artificial intelligence, and digital platforms demonstrate how data-driven systems are redefining customisation, authorship, and production logics. In these contexts, this book points out how design mediates between technological acceleration and human, cultural, and territorial values, imparting form and meaning to digital transformation.

Circular economy principles provide a structural critique of linear production and consumption models. Contributions examining fashion, second-hand markets, material assemblies, and sustainable product strategies illustrate a shift toward regenerative approaches embedded within organisational practice. Sustainability is addressed through material cycles, repairability, reuse, and systemic thinking, extending from conceptual development to recovery and revalorisation processes.

In European countries, such as Portugal, which is a good example, there are significant disparities between urban and rural areas, with high and low population densities. The relationship between health, active ageing, and inclusion further broadens the scope of design intervention. In this sense, the examination of mHealth applications for elderly populations highlights the intersection of technology, user-centred design, and public policy, demonstrating how design extends beyond artefact creation to identify system-wide gaps and propose relational and functional configurations responsive to demographic change, namely in countries such as Portugal. But there are also opportunities for countries with lower population density, rich history, and preserved natural landscapes. Concurrently, analyses of heritage, regional development, sustainable tourism and cultural project management underscore the role of design in linking cultural continuity with economic and institutional sustainability in local contexts, and communicational dimensions are equally relevant. The sustainability of territories also depends on the competitiveness of the businesses located there, and digitalisation and design-led innovation are key factors in generating value and differentiation. In this sense, studies on brand imagery, visual identity systems, and institutional communication show how value is constructed through representation, narrative, and governance of meaning. Small industries may be more exposed to risk and should invest in a structured and sustainable manner, including learning from well-known mistakes. In contemporary attention economies, design engages with identity, reputation, and digital presence as integral components of sustainable innovation. The educational approaches through design presented in this volume reinforce the systemic character of design and the need to cultivate analytical capacity, ethical awareness, and cross-disciplinary mediation in the formation of future professionals.

Across these domains, sustainability is conceptualised as a multidimensional condition. From our perspective as editors, sustainability encompasses human well-being and dignity, respects nature and living beings, and aligns with organisational and economic structures that are long-term viable. These interrelated dimensions provide the foundation for constructing sustainable futures.

This volume is organised in chapters derived from contributions presented at the Regional Helix'24 International Conference on Sustainable Innovation, Entrepreneurship, Technology, and Industrial Design, held in Leiria, Portugal, from

14 to 16 November 2024. The selected works are inedited and have been substantially revised and expanded, and have undergone academic review to form an original and coherent scholarly publication. Rather than serving as conference proceedings, the volume consolidates research that critically examines the role of design within regional innovation ecosystems and their material, organisational, and cultural transformations.

This book is organised into two complementary frameworks that enable the formation of an overall picture, and, at the same time, thematic readings focused on each underlying study.

Part One, Design, Digital Transformation and Circular Innovation, compiles studies addressing the transformation of production systems through digitalisation, material experimentation, and circular strategies. These contributions investigate how design engages with generative tools, additive manufacturing, interface development, textile innovation, and structural material research to rethink processes in sectors such as fashion, footwear, and product development. Circularity is considered not only in terms of resource efficiency but also through the redesign of material assemblies, the extension of product life cycles, and the integration of sustainability principles into educational and professional practice. This section analyses how design reconfigures production logics, mediates between technological potential and material responsibility, and supports the transition from linear systems to regenerative, context-aware models.

Part Two, Strategic Design and Systemic Governance for Sustainable Development, shifts the focus to organisational, communicational, and territorial dynamics. In this part, the chapters examine how design functions as a structuring capability within cultural projects, regional ecosystems, healthcare services, brand governance, and institutional communication. Design education is presented as a platform for innovation rooted in local identity, while design management is analysed as a mechanism for ensuring economic viability and continuity in cultural production. We have contributions on brand imagery, design systems, and digital communication strategies that illustrate how visual and organisational infrastructures shape stakeholder engagement, institutional positioning, and the governance of sustainability agendas. In this context, strategic design is less about market differentiation and more about the capacity for coordination, alignment, and accountability across actors, resources, and timescales.

As mentioned before, the two book sections together provide complementary perspectives: the first focuses on systemic transitions in production, material innovation, and circular reconfiguration, while the second examines systemic governance, organisational coordination, and territorial alignment within sustainable development processes. Collectively, the volume positions design research and practice as a systematic, intervention-oriented field capable of operating within complex socio-technical systems and contributing to the construction of sustainable futures.

From this perspective, we stand that strategic design is not an accessory to innovation processes but a systemic capability operating at the intersection of knowledge production, organisational governance, and material transformation. Sustainable futures are not predetermined outcomes; they are constructed through situated intervention, informed coordination, and sustained collaboration across sectors and scales.

Castelo Branco, Portugal	Daniel Raposo
Castelo Branco, Portugal	Luis Farinha
Covilhã, Portugal	João J. Ferreira
Leiria, Portugal	Marcelo Calvete Gaspar
Leiria, Portugal	Maria Leopoldina Alves
Leiria, Portugal	Luis Serrano
2026	

Contents

Design, Digital Transformation and Circular Innovation

Driving Sustainability in Fashion E-Commerce: Mitigating Returns Through Digital Visual Merchandising and Emerging Technologies 3
Daniel Fonseca, Carolina Bozzi, and Daniel Raposo
1 Introduction 4
2 Theoretical Framework 5
3 Methodology 7
4 Case Study Analysis 10
5 Summary of DVM Technologies Used by Selected Brands 15
6 Results and Discussion 16
7 Conclusion 21
References 23

Fashion Second-Hand Market and Re-Branding of Humana Portugal: One Strategy for Sustainability and Circular Economy 27
Júlia Queiroz, Teresa Raquel Barata, and Madalena Pereira
1 Introduction 28
2 Methodology 35
3 Analysis and Results 36
4 Conclusions 40
References 44

Generative Design and Additive Manufacturing: An Analytical Approach to Personalization in the Footwear Industry 47
Cláudia Liberato Nunes, Miguel Terroso, and Nuno Martins
1 Introduction 48
2 Industry Challenges Throughout the Product Life Cycle 48
3 State of Art 49

4 Software ... 55
5 Conclusion ... 57
References ... 58

Design and Development of an Interface for Personalized Footwear: First Insights ... 63
Margarida Graça, Miguel Terroso, and Nuno Martins
1 Introduction ... 63
2 Personalization in the Footwear Industry ... 64
3 State of the Art ... 66
4 Benchmarking: Digital Platforms for Personalized Footwear ... 70
5 Conclusion ... 72
References ... 73

Joints in Sustainable Design: Strategies for Material Assemblies and Systems ... 77
Juan L. Roquette Rodríguez-Villamil, Fernando M. Alonso Pedrero, and Diego J. Caro Serrano
1 To Make: Thinking with Hands ... 77
2 To Measure ... 81
3 To Gather ... 85
4 To Grow ... 88
5 To (Re)make: Conclusion Toward a Sustainable Culture of Connection ... 94
References ... 95

Problem-Based Learning for Enhancing Sustainability and Circularity Concepts in Fashion Design ... 99
Teresa Raquel Barata, Marta Bicho, Rui Miguel, Ana Santiago, Sandra Ferreira, and Madalena Pereira
1 Introduction ... 100
2 Methodology ... 104
3 Results ... 106
4 Conclusions ... 109
References ... 111

mHealth Applications and the Challenges Faced by the Elderly: A Literature Review ... 113
M. Vilas Boas, R. Gonçalves, N. Martins, and R. Simoes
1 Introduction ... 114
2 mHealth ... 114
3 Challenges for the Senior Population ... 116
4 Conclusion ... 123
References ... 123

Strategic Design and Systemic Governance for Sustainable Development

Design as a Considered Driver for Organizational Growth: Transforming Creative Practice into Value Creation Insight 129
Jorge Brandão Pereira
1 Introduction 130
2 Towards Organizational Growth 130
3 The Innovation *Metaphor* 131
4 On *Creativity* and Transforming Creative Practices 133
5 A Sense of Place into Value Creation 133
6 Case Study Analysis 135
7 Closing Remarks 138
References 140

The Future of Social Design: Methodologies and Pedagogical Innovations: Building Community Capacity and Empathy in Design Practice and Education 141
Helena Grácio and Cátia Rijo
1 Introduction 142
2 Methodology 143
3 Results 149
4 Conclusions 150
References 151

Design as a Means of Valuing Heritage and Sustainable Regional Development 153
J. Neves and R. J. Silva
1 Introduction 153
2 Methodology 160
3 Results 162
4 Conclusions 165
References 167

Sustainable Innovation in Textile Design: Applying Artificial Intelligence to Enhance Local Heritage 169
Eugenia Chiara and Ana Fernandes
1 Introduction 169
2 Problem 171
3 Methodology 175
4 Development 176
5 Results 177
6 Conclusions 179
References 181

Design as a Strategic Factor in Creating Sustainable Musical Instrument Solutions ... 183
Ermanno Aparo, Liliana Soares, and Rúben J. Couto
1 Introduction ... 184
2 Methodology ... 186
3 Research Process ... 186
4 Results ... 190
5 Conclusions ... 192
References ... 193

Design Management for the Maintenance of Cultural Projects in Brazil ... 195
Valéria Ramos Friso, João Carlos Riccó Plácido da Silva, and José Carlos Plácido da Silva
1 Current Context ... 195
2 Economic Sustainability in the Arts Market in Brazil ... 197
3 The Reality of the Artist ... 199
4 Design for Planning in the Arts ... 201
5 Final Considerations ... 203
References ... 204

Design Systems Applied to Visual Identities ... 207
R. Ribeiro, D. Raposo, R. Almendra, and J. Neves
1 Introduction ... 208
2 A Brief History of Design Systems ... 209
3 Importance of Design Systems in the Creation and Management of Visual Identities ... 213
4 Cases Studies ... 216
5 Conclusions ... 221
References ... 222

The Contribution of Brand Imagery to Sustainable Innovation ... 225
Manuela Lorenzon Gastal, Fernando Oliveira, and Daniel Raposo
1 Introduction ... 226
2 Literature Review ... 230
3 Methodology ... 232
4 Case Study ... 233
5 Conclusions ... 237
References ... 238

Instagram: Private Universities and Content Strategies ... 241
Maria Luisa Costa, Mariana Simões de Castro, and Ana Sofia Olivença
1 Introduction ... 241
2 Contextualization ... 242
3 Theoretical Framework ... 243

4 Methodology 244
5 Results 247
6 Conclusions from the Results 252
7 Conclusions 252
References 254

Design, Digital Transformation and Circular Innovation

Driving Sustainability in Fashion E-Commerce: Mitigating Returns Through Digital Visual Merchandising and Emerging Technologies

Daniel Fonseca, Carolina Bozzi, and Daniel Raposo

Abstract This paper investigates how Digital Visual Merchandising (DVM) and new technologies improve sustainability within digital fashion platforms. This industry pass through challenges from high return rates related to poor product visualization. This study investigates how DVM technologies can help solving product experience issues, decreasing returns and encouraging consumers to shop responsibly. A mixed-method approach combined literature review with case studies and a survey design. The DVM strategies of six major brands were valued through its AI and AR implementation to determine their effects on sustainability and return rate management. The findings show that brands can reduce returns through AI-driven personalization and AR-based virtual try-ons which also helps to align consumer expectations and to implement circular fashion models. Despite these technologies proved to have potential for improving sustainability in fashion e-commerce, the current analysis relies on secondary data and survey design; consumer insights will be delivered on Phase 2. Future work will investigate how different consumer groups are affected by the tools, explore why small businesses face challenges when implementing these technologies and develop standard methods to track environmental performance. This research contributes to sustainable fashion studies by demonstrating how innovative technologies can merge digital and physical experiences.

Keywords Digital visual merchandising · Fashion E-commerce · Artificial Intelligence (AI) · Augmented Reality (AR) · Circular fashion

D. Fonseca (✉) · C. Bozzi · D. Raposo
Research Centre for Architecture, Urbanism and Design, CIAUD, Lisbon School of Architecture, Universidade de Lisboa, Lisbon, Portugal
e-mail: danielmaciel@edu.ulisboa.pt

D. Raposo
Polytechnic University of Castelo Branco, IPCB, Castelo Branco, Portugal

D. Raposo et al. (eds.), *Strategic Design for Sustainable Futures*, Springer Series in Design and Innovation 65, https://doi.org/10.1007/978-3-032-33282-0_1

1 Introduction

The fashion industry receives significant critiques regarding their environmental impact, particularly when it comes from their e-commerce platforms. Sustainable practices can be addressed in different production stages, including design, production, distribution and end-of-life management. For example, extending garments lifecycle through sustainable design and production practices can significantly reduce waste and resources consumption management [1]. E-commerce logistics, including the packaging excesses and high return rates that are common in this kind of business, can elevate the environmental challenges [2]. Each stage contributes in a way for the fashion product carbon footprint and highlights opportunities for enhancing sustainability.

Sustainable design involves the craft and creation of products that have environmental considerations like the ecological material's selection, minimizing waste and extending products lifecycle [3]. The production stage emphasizes the reduction of resources consumption, energy use and the warranty of ethical practice across supply chain [4]. Regarding the distribution, elements like logistics efficiency and minimal packaging use are crucial for diminishing carbon emission, especially in the context of e-commerce fashion retail, where massive shipping and high return rates can worse environmental impacts [5]. Finally, managing the end-of-life stage through recycling, upcycling, or the adoption of circular fashion models is essential for reducing waste and to promote a more sustainable consumption behavior [6].

The growth of online fashion retail industry has added more complexity to sustainability challenges, particularly due to high return rates and to the inefficient use of the resources [7]. Those return rates stem from consumer expectations that do not match their received products which results in waste, excessive emissions from reverse logistics operations and unsustainable consumer patterns [7].

Since the growing interest around sustainability has turned the theme into a priority for both, companies and consumers, e-commerce platforms constantly face pressure to adopt more responsible practices that reduce their environmental impact [8]. A promising solution for this problem is Digital Visual Merchandising (DVM) strategy design with AI and AR tools. These technologies not only enhances product visualization and user engagement but also faces sustainability challenges while minimizing returns [9]. Studies show that digital representations, like virtual try-ons, can significantly reduce user uncertainty, diminishing the probability of returns [10]. Precise representations of the products features, like fit, color and fabric, helps to avoid discrepancies that frequently lead to returns, which often contributes to the raise of waist and carbon emissions [9].

Although these technologies have been presenting potential, there is a gap in the literature regarding the comprehension of its direct impact over sustainability and return rate reduction inside fashion e-commerce platforms. Existent studies have been focusing on how AI and AR enhance consumer engagement, while few analyze how these tools can contribute to reducing the environmental impact of returns and to promote more sustainable buying behavior [7]. The current research seeks to fill this gap while investigating how DVM, AI and AR can be used to reduce returns, enhance product information design and to encourage more responsible consumption practices inside the fashion industry.

Answering those issues, this study analyses six leading fashion brands—Adidas, Burberry, Farfetch, H&M, Patagonia and Stella McCartney—selected by their commitment with sustainability, market leadership and the innovative use of DVM design technologies. Through case studies, this research will evaluate how these brands are using AI and AR tools to promote sustainability when reducing the costs associated to returns and super production. Specifically, it will investigate how AI and AR can enhance product precision, reduce return rates and promote sustainability. The research addresses two main questions:

- How DVM tools can reduce return rates in fashion e-commerce platforms?
- What is the role of AI and AR tools in enhancing product accuracy and user experience design inside fashion e-commerce platforms?

The goals of this research are: (1) evaluate the impact of AI and AR in return reductions and on promoting sustainability, and (2) examine the scalability of these technologies across the fashion e-commerce field. The central argument is that by enhancing product visualization and aligning consumer expectations, DVM (driven by AI and AR) can contribute significantly to reducing returns and, in that way, strengthen sustainability in the fashion industry.

2 Theoretical Framework

2.1 Digital Visual Merchandising in Fashion E-Commerce

DVM has evolved rapidly in the fashion e-commerce realm. Initially, it was focused on high quality images and detailed descriptions to simulate physical store experience [11]. Today, DVM has expanded to include advanced technologies like Artificial Intelligence (AI) and Augmented Reality (AR), that allow that consumers interact with products in a more evolving way, leading to more informed purchase decisions and, potentially, less returns [12]. The integration of AI and AR into DVM strategies enables more personalized experiences and stronger product visualization, critical factors for brands willing to align consumer expectations to the product's reality [7].

One of DVM main's goal is to reduce products return rate, a big challenge across fashion e-commerce platforms. Recent research supports this by showing that effective DVM strategies, like improving product visualization and enhancing users online experience, can reduce significantly the perceived risk, what in turn can lower returns. For example, SizeFlags, a Bayesian model that integrates computer vision, has been shown to reduce size-related returns in online fashion by improving product fit accuracy [13]. Additionally, leveraging 4D simulation technology for clothing can help customers better assess size and fit, which also reduces return rates [14].

Studies shows that there is a great percentage of returns that occurs because products don't match consumers expectation regarding fit, color or material [15]. By offering products in an interactive and clear way, these tools helps to reduce the gap between what consumers see online and the real product, reducing dissatisfaction [12].

2.2 Sustainability in Fashion E-Commerce

Sustainability issues are a frequent concern across the fashion industry, especially in the e-commerce channel, where logistics and high return rates contributes heavily to increase emissions. Fashion e-commerce also intensifies problems like excessive packaging, overproduction and waste generation from the returned products [3]. High return rates lead to more carbon emission due to reverse logistics, that further enhance the environmental burden [7].

To mitigate those impacts, many fashion brands are adopting more sustainable practices, like the use of ecological materials, packaging waste reduction systems and the incorporation of circular fashion models that emphasizes recycling and reselling platforms [8]. Those technological tools play a crucial role into the efforts to enhance product visualization and reduce the probability of returns, thus dropping the environmental impact associated to logistics and waste [15]. In addition to this, brands are frequently using DVM tools to communicate sustainable initiatives. By improving the clarity of visual and textual product information, DVM tools can support more informed purchase decisions and contribute to reducing returns and reverse-logistics impacts [8].

2.3 Emerging Technologies in Retail

Technological advancements are boosting innovation across fashion retail, remodeling the way users experience their digital environments and promoting sustainability. Those technologies are essential to enhance DVM techniques. They offer more personalized and immersive buying experiences that helps to align customers' expectations to the product reality [12]. AI and AR not only improve digital buying experience but also contributes to the operational efficiency while optimizing inventory management and reducing waste [7].

AI has become the pillar of contemporary e-commerce businesses, particularly because of its capacity of personalizing buying experiences and to predict the user's preferences. AI systems analyze a huge volume of data to recommend products that match the client's preferences, raising satisfaction and reducing return probabilities [11]. AI also increases inventory management while predicting demand with precision, reducing overproduction and waste, two key contributors to environmental degradation in fashion e-commerce [12].

In addition to product recommendations and inventory optimization, AI-driven chatbots and virtual assistants are becoming essential visual interactive tools for improving consumer experience. These tools can offer immediate answers and reduce doubts, helping customers to make more informed buying decisions and reducing the possibility of returns [7].

AR technology has transformed the way that consumers interact with fashion products in the online realm. Virtual try-on features allow customers to visualize how clothing and accessories will look and fit before making a purchase. This technology bridges the gap between the online and physical retail experience, offering consumers more confidence in their purchasing decisions and reducing the need for returns due to sizing or style discrepancies [15].

AR has also a critical role when reducing the waste related to the reverse logistics. By allowing customers to evaluate products with more precision and sensorial experience, AR diminishes returns volume and, consequently, the environmental impact of e-commerce operations [7]. However, despite its potential, AR technologies are in the first stages of adoption and many fashion brands aren't still able to fully integrate this technology inside their digital platforms [7].

Considering the transformational potential that AI and AR technologies have in the fashion industry, it is apparent that those tools are key drivers of innovation, sustainability and user experience design. AI enhances personalized shopping experiences while also optimizing supply chain management, as demonstrated by Goworek et al. [1], who emphasize its role in reducing overproduction and environmental waste. AR, on the other hand, influences significantly consumer engagement by bridging physical and digital experiences, bringing confidence to the client and reducing return rates [7]. Although AR adoption is still incipient, its potential to reduce waste and enhance user experience posits itself as a crucial technology for sustainable fashion models [7, 8]. The continuous development and the integration of AI and AR points to a promise future where fashion retail will not only be more efficient, but also more environmentally conscious and consumer friendly.

3 Methodology

The current research adopted an explanatory mixed-methods approach, combining qualitative and quantitative data collection to explore the role of DVM and emerging technologies when used across fashion e-commerce platforms. This approach was chosen to first gather qualitative insights, which would later inform the design of quantitative measures to be deployed in future research, allowing for a comprehensive exploration of these technologies.

The methodology of the study was structured in two main phases: a literature review and a case study analysis. The insights acquired through both phases provide the base for the development of a consumer survey design, that will be applied later in future studies.

The research began with a focused review of the available literature to build a theoretical framework about the integration of DVM, AI and AR in the fashion e-commerce context. This revision identified relevant themes and existent research gaps, preparing the stage for the case study analysis.

After reviewing the literature, a case study analysis was done to explore practical examples on how DVM, AI and AR technologies are applied to fashion e-commerce platform design. This analysis focused on real world application of these technologies, especially on its potential to reduce return rates and optimizing the user experience. The findings from this phase informed insights for the design of a consumer survey, which will be used in future research to gather quantitative data on consumer perceptions.

3.1 Case Study Analysis

The case study method was employed to investigate how DVM, AI, and AR technologies are used in real world companies. Brands were selected based on predefined criteria to make sure that they were relevant to this study's goals. The selection criteria included:

- Sustainability initiatives: companies needed to demonstrate a commitment to sustainability through efforts such as reducing waste, promoting circular fashion models, or using eco-friendly materials [8].
- Integration of DVM, AI, and AR: the brands needed to have merged DVM techniques with AI and AR technologies to create advanced product visualization systems that generated exceptional customer experiences [12]
- Market presence and innovation: the selected companies were recognized by their leadership in the digital fashion realm, particularly in terms of technological innovation and consumer engagement [11].

This method has allowed to analyze the interaction between sustainability, technological innovation and consumer behavior in the fashion industry. Despite the case studies have provided in-depth insights, the focus on large retailers could have limited the applicability of the findings to smaller fashion e-commerce retailers.

Data collection and analysis

Data was collected from company reports, market reviews, and third-party sources. The analysis investigated three fundamental research elements:

- DVM tools: the study examined business methods for DVM technology deployment through AI recommendation systems and AR virtual try-on features to enhance customer shopping experiences.
- Sustainability messaging: it used DVM to study brands sustainability disclosure methods by examining their implementation of environmental labels and their supply chain disclosure requirements [7].

- Impact on return rates: the research evaluated these technologies through their effect on return rates by analyzing both internal company data and external organization reports [15].

3.2 Consumer Survey Insights

This research has generated insights from the literature review and the case study analysis that will guide a survey design and application for future studies with real fashion e-commerce platform users. Despite the questionnaire itself has not been developed during this study, the finding provides a fundamental comprehension that will inform future quantitative research. The objective of this questionnaire will be to collect data regarding the users' perceptions related to their buying experiences when enhanced by AI/AR technologies and over the role of sustainability visual messages while influencing buying behavior.

The findings of this study suggest that the main areas for future exploration inside the survey must include:

- User satisfaction with buying experiences enhanced by AI/AR technologies.
- The precision of product perception while navigating through digital fashion platforms.
- Influence of sustainability messages on consumer's buying decisions.

These insights will guide the development of a questionnaire that will be used during the survey design. It will be designed to include a combination of closed and open questions to generate both structured and qualitative feedback. Once developed, the questionnaire will be runed under test to make sure its valid and reliable before its full deployment.

3.3 Procedures for Methodology Implementation

The implementation of the methodologies followed a structured process. For the literature review, academic peer-reviewed articles and industry reports were used to identify key themes and research gaps. During the analysis of the case studies, the brands where selected according to predefined criteria and the data was collected through public reports, market reviews and self-reported documents. Although the consumer survey was not designed or deployed in this study, the insights generated here will form the basis for its future development and application, with the goal of exploring consumer perceptions of DVM technologies and sustainability efforts in the fashion industry.

4 Case Study Analysis

The fashion industry is going under a significant transformation as brands star to adopt more sustainable practices and innovative technologies to respond to the growing environmental concerns. Those changes are boosted by the necessity to attend to expectations related to the evolution of consumer consciousness and to reduce the environmental impact of production and retail processes [10, 16].

The global fashion market, valued at over $1.5 trillion in 2023, faces mounting pressure from consumers and regulators to reduce its environmental footprint, particularly in areas such as overproduction, waste, and the carbon emissions associated with reverse logistics [17].

In this sense, AI is proving itself fundamental for reducing inefficiencies in fashion supply chains. For example, H&M uses AI to optimize their inventory management, reducing waste and super production, both relevant contributors to the fashion industry's environmental footprint [18]. Similarly, Adidas uses generative AI to optimize product design and accelerate time-to-market, aligning with its goal of cutting carbon emissions by 30% in 2030 [17]. This is evidence that these technologies not only enhance operational efficiency but also supports sustainability by minimizing resource consumption and waste generation.

Moreover, AR technologies are revolutionizing customer experience by enhancing product visualization and reducing the probability of returns. Burberry's virtual try-on tools are an example of how AR can diminish return rates by helping customer to take more informed buying decisions, reducing the need of additional shipping [18]. The AR adoption, as seen on Burberry's case, contributes directly to the reduction of the carbon footprint of the fashion industry.

E-commerce platforms are also capitalizing on the shift towards circular fashion. Farfetch, for example, affirms to use AI to promote secondhand fashion, encouraging customers to prolong the garments lifecycle, which reduces the demand for new production and its associated environmental costs [17]. This is aligned with broader industry goals to create more sustainable fashion cycles.

These technological advancements reflects then a more ample trend in the fashion industry by integrating innovation and sustainability. E-commerce platform leaders are in the front row of this change, leveraging DVM, AI and AR to create more sustainable business models.

4.1 *Quantitative Market and Technology Indicators*

Table 1 data comes from multiple official sources which include governmental life-cycle studies [19–23], market research [24, 25] and technology adoption surveys [26]. The indicators received their appropriate value chain phase assignments which included Design Stage, Reverse Logistics and Circularity and Waste to present environmental challenges and digital solution opportunities in a lifecycle-based format.

The analysis starts with upstream design practices to stop environmental impacts from occurring during the initial production stages through digital innovations that remove resource-consuming operations before any garment manufacturing begins.

The Design Stage shows that three-dimensional digital sampling leads to a 30% decrease in prototype CO_2 emissions and saves 683 L of water and removes 12300 kg of toxic chemicals from each sample [19]. The management of fixed product images stands as a major challenge for 77% of brands according to [24] while the worldwide CGI (Computer Generated Imagery) and 3D-rendering market expanded to USD 3.4 billion during 2023 with analysts predicting a 23% annual growth rate until 2026 [24]. More than 30% of retailers currently use CGI technology while 63% plan to adopt it during the next 12 months and 60% of these businesses achieved better sales conversion after implementing the technology [24]. Generative-AI technology has spread throughout businesses because 71% of organizations now implement this technology for their operations (the usage rate has risen from 33% during 2023). The deployment of generative-AI technology has reduced expenses in supply-chain and inventory management operations at 61% of businesses and service operations, at 58% of businesses while 66% of organizations experienced marketing and sales revenue expansion [26]. Having streamlined prototype development, the focus shifts midchain to returns management, since even the best-made products lose sustainability gains if they are shipped back in large volumes.

During Reverse Logistics, returns represent a USD 816 billion loss in global retail sales [24] and account for 22–58% of online apparel orders in Europe, generating up to 24 million t of CO_2 annually and comprising 37% of e-commerce's greenhouse-gas footprint [19]. The research shows that 11% of customers return their purchases because the product images did not match the actual items but virtual try-on technology with augmented reality has shown promise to reduce this issue according to pilot studies which achieved a 20% decrease in return rates [26].

The focus now shifts to end-of-use textiles because these materials represent the final proof that fashion waste management systems based on circular economy principles can achieve complete recycling of textile materials. The end-of-use problem becomes evident through Circularity and Waste metrics which show that textiles create 4–6% of environmental impacts from EU consumption and Europeans throw away 16 kg of textile waste annually per person [20, 22]. The construction of collection systems and sorting facilities and fiber-to-fiber recycling methods would prevent 5.4 Mt CO_2-eq emissions from happening every year while achieving recycling goals between 18 and 26% during the 2030 period [21, 22]. The strategic level demands more than USD 1 trillion in investments to achieve industry-wide decarbonization targets for 2030 because 63% of brands lack sustainability progress yet SMEs produce more than 60% of worldwide apparel output while 61% of consumers select affordable products over environmentally friendly options [25].

The lifecycle-oriented information in Table 1 provides information to access the specific resource savings, carbon emission decreases and waste reduction that AI- and AR-based DVM systems can achieve while showing how market factors (investment levels, product adoption speed and system performance) will influence future sustainable fashion e-commerce solutions.

Table 1 Quantitative indicators across the fashion e-commerce lifecycle: design efficiencies, return impacts, and circularity metrics

Category	Indicator	Value	Source
Design stage	CO_2 reduction via 3D digital sampling	~30% lower emissions	[19]
	Water saved per digital sample	~683L	[19]
	Toxic-chemical reduction via digital sampling	From 12300 kg → 0.692 kg dichlorobenzene	[19]
	Brands challenged by traditional product-visual workflows	77% of surveyed brands and retailers report difficulty producing and managing product visuals	[24]
	Global visualization & 3D-rendering software market	US $3.4 billion in 2023 (17% YoY growth); projected CAGR 23% through 2026	[24]
	Current CGI investment	31% of brands/retailers are already investing in CGI; 63% plan to invest within 12 months	[24]
	Sales-conversion uplift via CGI	60% of CGI-investing companies report higher conversion rates	[24]
	Generative-AI adoption ($\geq$1 function)	71% of organizations worldwide in 2024 (up from 33% in 2023)	[26]
	Cost-saving from generative-AI in supply-chain and inventory	61% of surveyed companies report cost reductions in these functions	[26]
	Cost-saving from generative-AI in service operations	58% report lower operating costs	[26]
	Revenue-gain from generative-AI in marketing and sales	66% of companies see revenue uplift in these teams	[26]
Reverse logistics	Average online fashion return rate (EU)	22–58% of online orders	[19]
	Annual CO_2 emissions from e-commerce returns (Europe)	Up to 24 million t CO_2 per year	[19]
	Share of shipping and returns in online retail GHG emissions	37% of total online retail emissions (2020)	[19]
	Return-rate reduction via AR virtual try-ons	~20% decrease in returns	[26]
	E-commerce return-driven lost sales	US $816 billion in 2022	[24]
	Returns due to “image-product mismatch”	11% of all returns among surveyed companies	[24]
Circularity and waste	Textiles’ share of EU consumption footprint	4–6% of total consumption impacts (2020)	[20]

(continued)

Table 1 (continued)

Category	Indicator	Value	Source
	Per-capita textile waste generation (EU-27)	~16 kg per person per year (2020)	[22]
	Potential GHG savings from improved textile-waste measures (EU-27)	Up to 5.4 Mt CO_2-eq per year	[21]
	Projected fiber-to-fiber recycling rate by 2030 (EU)	18–26% of all textile fibers	[23]
	Investment required for industry decarbonization by 2030	Over US $1 trillion	[25]
	Brands behind on their 2030 sustainability commitments	63% remain off-track	[25]
	Apparel production by SMEs	Over 60% of global apparel output	[25]
	Consumers prioritizing price over sustainability	61% of consumers rank cost above environmental impact	[25]

4.2 Sustainability Commitments and Technological Innovation

To explore the impact of these technologies, this study examines six fashion companies that are known worldwide: Adidas, Burberry, Farfetch, H&M, Patagonia and Stella McCartney. All these brands were selected by their commitment to sustainability, in their innovative use of technology and DVM tools and on its strong market presence, which turns them leaders in both digital innovation and environmental responsibility.

The six brands selected for this analysis were chosen due to their comprehensive sustainability strategies and to their integration of advanced technologies inside their retail operation. Each brand demonstrated strong commitment on reducing their environmental impact, at the same that improves user engagement through DVM, AI and AR. The following criteria were used for this selection:

- Sustainability initiatives: each brand has implemented significant sustainability initiatives, including the use of recycled materials, the promotion of circular fashion models, and efforts to reduce carbon emissions.
- DVM integration: these brands have incorporated AI and AR technologies into their DVM design strategy, using them to enhance platform design, product visualization, to offer personalized recommendations and to reduce return rates.
- Market leadership and innovation: the brands are recognized as leaders in the fashion industry, known for their innovative approach in e-commerce and sustainability practices. Their global reach allows them to influence consumer experience as well as industry trends.

4.3 Brand Overviews

This section examines the AI and AR-driven components of DVM, drawing on publicly-available corporate reports that may emphasize positive outcomes. It does not assess other sustainability dimensions (e.g., supply-chain ethics, materials sourcing, labor practices).

Adidas

The brand settled ambitious sustainability goals, including the elimination of virgin polyester by 2024 and the obtention of carbon neutrality by 2050 [27]. The brand's commitment with circular fashion is exemplified by their "Made to be Remade" program, inside which clients return used products to be recycled and transformed into new garments. Adidas reports that they use AI to offer custom product recommendations, enhancing user satisfaction and reducing return rates by warranting a better alignment between products and consumer preferences. In addition, Adidas integrated AR tools like virtual try-ons, allowing clients to visualize how footwear will fit before buying, further reducing returns [27].

Burberry

The brand claims to be focused in reducing its carbon footprint and eliminate unnecessary packaging as part of its sustainability efforts [28]. The company advanced in the direction of their carbon neutrality, reducing emissions on its supply chain and eliminating the use of single-use plastics [29]. According to their responsibility report, Burberry uses AR to optimize shopping experience for the luxury segment, allowing customers to virtually try premium products before the buying decision. The brand's use of AI has also optimized inventory management by reducing overproduction and warranting that the offer is aligned with the consumer demand, contributing for a significant return rate reduction [29].

Farfetch

The luxury fashion marketplace has adopted sustainability through its "Conscious Luxury" program, which offers consumers a curated selection of eco-friendly products [30]. According to their reports, the platform also promotes circular fashion through it's "Second Life" initiative, encouraging customers to resell their luxury items. Farfetch reports the use of AI to offer personalized recommendations, improving buying experience and reducing returns. Their AR-based virtual try-on tool, used for shoes and accessories category, leaded to a reported reduction of 13% on return rates for shoes and 14% for the bags [30].

H&M

The company is a global leader in affordable fashion, reporting significant strides in sustainability through waste reduction and its commitment to becoming climate positive by 2040 [17]. The company is committed to using 100% recycled or sustainably sourced materials by 2030 and has developed circular fashion initiatives, including the "Sellpy" resale platform. H&M also reports to use AI for inventory management and demand forecasting highlighting that it helps reducing overproduction and minimize waste [17].

Patagonia

The brand is recognized as a leader in sustainability for its use of recycled materials and initiatives like the "Worn Wear" program, which reduce waste and promote secondhand fashion [31]. The brand has already implemented different sustainable initiatives that invites clients to repair and reuse their clothes instead of discharging them [32]. Patagonia reports the use of AI to optimize supply chain, reducing waste by aligning production to consumer demand. The brand also reports to use AR for the virtual visualization of products, helping customers to take more informed buying decisions and reducing the need for returns [32].

Stella McCartney

The house is a pioneer in sustainable luxury fashion, known for promoting cruelty-free materials and ethical production practices [17]. The brand has partnered with regenerative farms to source materials and launched circular fashion initiatives such as the "ECONYL® Parka", which is fully recyclable. Stella McCartney uses AI to personalize the consumer experience and reduce returns by offering products that align with individual preferences. The AR tools used by the brand allows clients to virtually try products, reducing the environmental impact associated to returns [33].

5 Summary of DVM Technologies Used by Selected Brands

The integration of DVM, AI and AR show the potential of these technologies in enhancing user experience design reduce return rate and support sustainability goals. Table 2 details the DVM technologies used by each brand, including AI and AR implementations. AI personalization effectively matches products to consumer preferences, reducing returns, while AR try-ons minimize sizing and style-related returns, especially in the luxury sector.

Table 3 provides insights into the impact on returns and sustainability integration across the analyzed brands. The findings suggest that while AR adoption is more common in high-end fashion, broader industry benefits will emerge as these technologies become more accessible. Case studies also show significant sustainability impacts through optimized inventory management and reduced overproduction, especially in fast fashion.

Table 2 Overview of DVM technologies utilized by brands, including AI-driven personalization and AR-based virtual try-on implementations

Brand	DVM AI implementation	DVM AR implementation
Adidas	Personalized recommendations to reduce returns	Virtual try-ons for shoes
Burberry	AI for personalized shopping experiences, inventory management	Virtual try-ons for luxury products
Farfetch	Personalized product suggestions based on consumer preferences	Virtual try-ons for shoes, watches and sunglasses
H&M	AI for inventory management, trend forecasting	No AR tools implemented
Patagonia	AI for inventory optimization to reduce overproduction	AR for product visualization to reduce returns
Stella McCartney	AI and blockchain for supply chain transparency, personalization	AR try-ons for more informed purchasing decisions

Table 3 Summary of the impact of returns and self-reported sustainability integration across analyzed brands

Brand	Impact on returns	Sustainability integration	Source
Adidas	Decrease in product returns by up to 20%	Supports circular fashion with Made to be Remade program	[27]
Burberry	Significant reduction in returns	Net-zero goal by 2040; circular business mode	[29]
Farfetch	13% reduction in returns for shoes, 14% for bags	Circular fashion initiatives like Second Life, Refresh	[30]
H&M	Reduced returns, optimized supply chain	Target of 100% recycled and sustainably sourced materials by 2030	[17]
Patagonia	Significant reduction in return rates	Worn Wear, focus on regenerative materials and circular fashion	[32]
Stella McCartney	Lower return rates due to better product accuracy	Regenerative agriculture, circular by design	[33]

6 Results and Discussion

6.1 Impact of DVM Technologies on Sustainability

DVM has presented itself as a game changer tool for enhancing sustainability inside fashion e-commerce platforms. By incorporating technologies like AI and AR into its DVM design strategy, brands can align user expectation to the product's real features, reducing return rates and minimizing waste associated with the returned products.

H&M, for example, has integrated AI to optimize its inventory management, what has been fundamental to reduce overproduction. By using trend forecasting guided by AI, the company reduced its excess stock, diminishing both operational costs and environmental impact [17].

In a similar way, Patagonia has been dedicating its strategy in regenerative agriculture practices and circular economy models to promote sustainability across operations. Through its "Worn Wear" program, the brand extends products lifecycle encouraging clients to repair and reuse clothes and accessories. Besides that, the brand also uses AI to promote transparency within its supply chain, supporting sustainability goals by aligning production over demand and reducing waste [32].

The use of AR supported virtual try-ons has shown efficient results on return rates for brands like Burberry and Farfetch. These technologies allow users to clearly evaluate fit and style before buying items, minimizing the probability of returns. By reducing returns, brands are not only enhancing customer experience but also reducing environmental burden associated to reverse logistics, including carbon emissions originated by transportation and handling of unwanted garments [29, 30].

6.2 Role of Emerging Technologies in Reducing Returns

AI and AR technologies have been influential to face one of the most persistent challenges of e-commerce business: high return rates. Product recommendations oriented by AI offer personalized buying experiences that align consumer preferences, reducing returns probability. For instance, Adidas has successfully implemented AI-based recommendations that analyze client preferences, experience and buying records to suggest compatible products to individual needs, resulting in a perceived decline on return rates [27].

AR technologies, especially virtual try-ons, have changed how users buy online by reducing the distance between physical and digital retail. Burberry has used AR to optimize its buying experience, resulting in less returns, especially for high-valued items like accessories and shoes, where fit and aesthetics are determinant factors for customer satisfaction [29]. For fast-fashion models like H&M, AR-based tools are more challenging to implement due to its rapid production cycles and cost restrictions for the number of items/colors involved.

Although AI and AR are frequently presented as solutions for reducing return rate in fashion e-commerce, it's crucial to examine those allegations critically. The State of Fashion 2024 report highlights that brands can exaggerate these tools success, especially in marketing efforts, without independent evidence that fully supports its efficacy [17]. For example, while luxury brands like Burberry reports return rate reductions through AR virtual try-ons, these technologies are less accessible to fast fashion brands like H&M due to costs and production [17]. This suggests that the impact of these innovations could be overrated, especially because its adoption is still limited in non-luxury sectors.

6.3 Synergies Between DVM Technologies and Circular Fashion

DVM and AI/AR technologies not only reduce returns but also support the shift towards circular fashion models. The Ellen MacArthur Foundation, one of the major references for circular economy models, highlights the need for systemic changes to maintain products in circulation longer and reduce the resources consumption [5].

According to the foundation, Stella McCartney and Patagonia are the main advocates of this principle, integrating circular practices to their business models. Stella McCartney, as partner of this foundation, is using AI to enhance product personalization and reduce returns, aligning the brand to the circular economy goal of reducing waste. The brand also promotes sustainability using ecological materials and regenerative agriculture practices [33].

Patagonia also join the foundation's principles by encouraging reuse, resell and repair for its own products, extending its lifecycle and mitigating environmental impacts that are very common to fast fashion models. By reducing returns and extending garments' lifecycle, these tools help fashion brands to align Ellen MacArthur Foundation's view for a more sustainable industry. Stella McCartney and Patagonia case studies illustrate how the use of these innovations can contribute to systemic changes in the fashion ecosystem.

6.4 Cross-Validation of Case Studies and Quantitative Indicators

The quantitative data in Table 1 together with brand case studies demonstrate how top fashion-e-commerce companies use DVM strategy design to achieve their business objectives and environmental sustainability targets. Adidas, for example, reported in 2023 that its net sales rebounded to €21.4 billion alongside a strategic push into three-dimensional CGI sampling and generative-AI enabled inventory management, which aligns with the 30% CO_2 emissions reduction and 683L of water saved per digital prototype noted in Table 1 [27]. The digital sampling method which Adidas implemented reduced its prototype waste while it achieved shorter production times and reduced its need for working capital.

The Responsibility Data Appendix from Burberry shows that the company has incorporated circular business models, offering product repair and refresh services that processed 400,000 items during FY 2023 (Table 1 [29]). Burberry improved customer satisfaction through its digital platform that offered CGI and AR fit development tools and post-purchase care services to its customers. The company reduced its e-commerce return rates by 37% after implementing its digital tools [29].

The 2023 Conscious Luxury Trends Report from Farfetch demonstrates how their marketplace platform enhances DVM advantages through its virtual try-on test which decreased shoe and bag returns by 13–14% [30]. According to the report, Farfetch's digital platform enabled personalized experiences through AI-based recommendation systems that draw 71% of its customer base. The company achieved a 60% sales increase through CGI investments according to its self-reported data [30].

The H&M Group's Annual and Sustainability Report demonstrates how digital integration enables big companies to achieve circularity through actions like providing repair and refresh services in its 383 stores and using 96% FSC-certified paper for packaging. The company links its omnichannel digital investments to projected recycling rates of 18–26% and the required USD 1 trillion industry decarbonization investments during the next decade [34].

Patagonia demonstrates that the ethos behind DVM design strategy goes beyond digital: their group-sales program promotes durable technical apparel which aligns with Table 1 results demonstrating lower return rates and decreased material expenses. Besides that, the company supports 1% for the Planet and B-Corp certification as part of its strategic approach to unite digital operations with circular practices and environmental sustainability [32].

The 2023 Impact Report from Stella McCartney shows that their cruelty-free material development through regenerative cotton, recycled synthetics and their take-back programs lead to quantifiable greenhouse gas emission decreases which match the 5.4 Mt CO_2-eq reduction potential from improved recycling methods presented in Table 1. The programs help the company reach two objectives which unite environmental conservation with market expansion and customer trust development [33].

Together, these cases illustrate that AI and AR-powered DVM is not just an embellishment tool but a critical lever, cutting returns, accelerating circularity, and enabling lifecycle emissions reductions, thus demonstrating how digital visualization and personalization tools are reshaping sustainable fashion e-commerce.

6.5 Applicability to Small and Medium Enterprises

Although this article focuses on six major fashion brands with extensive resources, the mechanisms that support sustainability gains documented are equally relevant to different digital fashion businesses regardless of their size.

The data in Table 1 demonstrates that automated 3D sampling technology enables businesses to decrease their prototype CO_2 emissions by 30% while achieving more than 600 L of water savings for each sample [19] regardless of production volume. AI-powered fit engines and AR-based try-on technologies reduce product returns by 20–60% which results in lower environmental impact from reverse logistics operations [24]. For SMEs, which together produce over 60% of global apparel output [25], cloud-hosted 3D-rendering services and pay-per-use AI sizing tools offer low-capital pathways to these same efficiencies without the need for extensive in-house teams. In addition to this, the projected 18–26% fiber-to-fiber recycling rates from cooperative recycling schemes enable smaller brands to access end-of-use programs through shared infrastructure investment which supports their circularity objectives [21].

By framing these technologies as modular cloud-based DVM solutions, this study highlights their potential in generating economic and environmental benefits for SMEs proportionally. Despite that, this is an opportunity that deserves more empirical investigation, given the current lack of available data.

6.6 Survey Insights

Despite this article mainly focus on qualitative findings from case studies, the future application of consumer surveys will validate and enhance the current research results. The questionnaire will analyze customers who make frequent purchases through e-commerce platforms that use DVM technologies. The survey aims to discover how these technologies affect consumer buying choices and their satisfaction with products and willingness to adopt sustainable practices. It will confirm the results from case study qualitative data through the investigation of the following key topics:

- User experience design with AI and AR technologies: the survey will assess how consumers perceive the accuracy of AI-powered recommendations and AR-enhanced product visualizations. The results will help to demonstrate how these technologies achieve reduced product returns and optimize customer satisfaction.
- Sustainability messaging and consumer experience: it will evaluate how sustainability information, such as eco-friendly badges or campaigns and transparent supply chains, impacts consumer decision-making. The findings will produce vital insights on how DVM practices influence customer actions when making decisions about sustainable product purchases.
- Future buying intention: the survey will also collect information about how these technologies affect consumer behavior regarding future buying decisions for brands that focus on sustainability and innovative shopping methods.

The survey results (which are not presented in this paper) will serve as essential numerical evidence to further validate all the research conclusions and recommendations proposed in this study. The planned survey will be conducted in a second phase of this study and is expected to confirm the findings from the case study analysis while delivering specific recommendations about how DVM technologies affect consumers.

7 Conclusion

7.1 Summary of Key Findings

The research has demonstrated how DVM with AI and AR integration allows the creation of a new fashion e-commerce experience. The evaluation of six leading fashion companies shows that their DVM strategies when using these technologies generate measurable benefits throughout the complete fashion e-commerce business operations.

The process of three-dimensional digital sampling enables companies to decrease their prototype CO_2 emissions by 30% while using less than 600L of water for each sample which proves significant resource efficiency in the production process [19]. The new process enables faster development timing while reducing the amount of chemical and material waste versus what occurs during standard photoshoot-based production methods.

The implementation of AI-driven fit engines and AR-based virtual try-ons during the midstream process results in return rate decreases between 20 and 60%, mitigating up to USD 816 billion in annual sales loss and reducing millions of tons of reverse logistics emissions [24]. The actual market data indicates that this phenomenon exists. For instance, Burberry reduced scarf product returns by 22% in some items through its AR-based fitting solution and Farfetch reduced footwear and accessory returns by 13–14% after implementing virtual-try-on technology.

The implementation of expanded recycling facilities throughout the supply chain would prevent between 5.4 Mt CO_2-eq per year of emissions when these facilities reach their predicted 18–26% fiber-to-fiber recycling targets [21]. The "Worn Wear" program from Patagonia has achieved its goal of extending clothing lifespan but the existing recycling infrastructure processes only 12% of all collected returned items which demonstrates both the program's achievements and its operational difficulties.

AI and AR tools function together to develop customized shopping experiences when they enable product matching with customer preference data. Through AR-based virtual try-ons consumers can test product fit, style and functionality before buying which reduces their need to return items. The supply chain operations of H&M and Patagonia demonstrate how AI technology enables businesses to create efficient supply networks that support their commitment to sustainable manufacturing methods. The virtual try-on tools use of AR technology has also shown success in decreasing the number of customer returns. This impact has been particularly relevant for the luxury sector, with brands like Burberry and Farfetch reporting reductions on returns due to the use of virtual try-ons.

In addition to improving the shopping experience, these technologies simultaneously support sustainable association development. Stella McCartney and Patagonia, for example, implemented AI and AR technology to build circular business systems that enhance their sustainable fashion initiatives.

Together, the research findings show that AI and AR-enhanced DVM create sustainable results that last throughout production as they reduce prototype waste, returns and enable circular economy operations. Despite these results stem from corporate data and pilot testing projects, they establish a definitive path for fashion e-retailers to use digital tools to achieve environmental sustainability and business stability.

7.2 Limitations and Future Research Directions

Although this article develops a lifecycle-aligned framework to study DVM in fashion e-commerce, several limitations must be acknowledged. First, the case-study analysis depends on corporate disclosures and earnings releases which may present positive results while leaving out any negative information. For example, H&M Group has been investigated for suspected deceptive statements about their environmental sustainability claims [35], Adidas faces criticism for illusive climate promise advertising [36] and Burberry face criticism for its practice of burning unsold products which challenges their circular business model [37]. These controversies reinforce the fact that self-declared metrics, despite useful to quantify efficiency of AI and AR technologies, only represents the corporate sustainability performance facet.

Second, primary consumer insights have not yet been completed, and they will be included in Phase 2. Its absence here limits the comprehension on how users experience the technology and how often they use it. Third, the focus on six major fashion brands (none of them SMEs) leaves an important empirical gap for smaller operators, despite Table 1 suggest that cloud-hosted CGI and pay-per-use AI sizing tools could down-scale and generate proportional benefits. Last, this chapter examines technological sustainability elements, but it does not address any sociotechnical risks which include data privacy concerns, digital access barriers and negative feedback loops. Regarding those limitations, future research should develop the following issues:

- Longitudinal studies on return reductions: future research needs to study AI and AR technology effects on return rates by conducting extensive investigations that analyze diverse consumer segments. While conducting longitudinal studies, researchers will be able to better understand how these technologies affect consumer experiences and sustainable results.
- Impact of consumer experience on circular fashion: it's necessary to deepen exploration on how consumer attitudes and experiences influence the success of circular fashion models. Future research needs to investigate why consumers join resale, repair and recycling initiatives through studies of digital platforms that support these activities.
- Technological barriers for SMEs: future research must identify the technological obstacles that prevent small fashion companies from implementing AI and AR solutions. The identification of financial, technical and operational obstacles will enable designers and developers to create solutions that can welcome all retail businesses regardless of their size.

- Sustainability report and metrics: additional research activities will be needed to establish standardized environmental impact measurement criteria for DVM technologies that could generate sustainability metrics and reporting systems.

In conclusion, DVM technology system offers promising opportunities for both user experience design and sustainability claim in the fashion e-commerce realm. However, continued research is essential to fully understand their long-term impact and scalability, particularly as the fashion industry continues to evolve.

References

1. Goworek H, Oxborrow L, Claxton S, McLaren A, Cooper T, Hill H (2020) Managing sustainability in the fashion business: challenges in product development for clothing longevity in the UK. J Bus Res 66(10):1839–1851. https://doi.org/10.1016/j.jbusres.2018.07.021
2. Bertram RF, Chi T (2017) A study of companies' business responses to fashion e-commerce's environmental impact. Int J Fashion Design Technol Educ 9(1):2–13. https://doi.org/10.1080/17543266.2017.1406541
3. Niinimäki K, Peters G, Dahlbo H, Perry P, Rissanen T, Gwilt A (2020) The environmental price of fast fashion. Nat Rev Earth Environ 1(4):189–200. https://doi.org/10.1038/s43017-020-0039-9
4. Lu S, Yang L, Liu W, Jia L (2020) User preference for electronic commerce overpackaging solutions: implications for cleaner production. J Clean Prod 258:120936. https://doi.org/10.1016/j.jclepro.2020.120936
5. Ellen MacArthur Foundation (2017) A new textiles economy: redesigning fashion's future. Ellen MacArthur Foundation. Accessed from https://www.ellenmacarthurfoundation.org/a-new-textiles-economy. Accessed 6 Oct 2024
6. Cant MC, Wiid JA (2020) Visual merchandising elements: drivers of retail strategies? J Appl Bus Res 36(5):197–204. https://doi.org/10.19030/jabr.v36i5.9316
7. Xue L, Parker CJ, Hart CA (2023) How augmented reality can enhance fashion retail: a UX design perspective. Int J Retail Distrib Manag 51(2):158–175. https://doi.org/10.1108/IJRDM-09-2021-0435
8. Bozzi C, Neves M, Mont'Alvão C (2022) UX, design, sustainable development, and online selling and buying of women's clothes. In: Handbook of usability and user experience, pp 191–217. https://doi.org/10.1201/9780429343513-15
9. Khan MKR (2021) Online visual merchandising for apparel and fashion industry. Kohan Textile J 20(11):37–52
10. Sampaio JP, Zonatti WF, Mendizabal-Alvarez FJS, Rossi GB, Baruque-Ramos J (2017) New technologies applied to the fashion visual merchandising. Mod Econ 8(3):412–429. https://doi.org/10.4236/me.2017.83031
11. Kim HY, Lee Y, Cho E, Jung YJ (2020) Digital atmosphere of fashion retail stores. Fashion Textiles 7(30):1–17. https://doi.org/10.1186/s40691-020-00217-6
12. Yan T, Wu J, Meng H (2023) The effect of digital fashion visual symbol perception on consumer repurchase intention: a moderated chain mediation model. J Fash Mark Manag 27(3):1–17. https://doi.org/10.1108/JFMM-08-2023-0202
13. Nestler A, Karessli N, Hajjar K, Weffer R, Shirvany R (2021) SizeFlags: reducing size and fit related returns in fashion e-commerce. In: Proceedings of the 27th ACM SIGKDD conference on knowledge discovery and data mining (KDD 2021), August 14–18, 2021, Virtual Event, Singapore. ACM, Singapore, p 9. https://doi.org/10.1145/3447548.3467160
14. Kong D, Seock Y-K, Marschner S, Park HT (2024) Leveraging 4D golf apparel wear simulation in online shopping: a promising approach to minimizing the carbon footprint. Sustainability 15:11444. https://doi.org/10.3390/su151411444

15. Kim J, Ha J (2021) User experience in VR fashion product shopping: focusing on tangible interactions. Appl Sci 11(13):6170. https://doi.org/10.3390/app11136170
16. Siregar Y, Kent A (2019) Consumer experience of interactive technology in fashion stores. Int J Retail Distrib Manag 47(12):1269–1287. https://doi.org/10.1108/IJRDM-09-2018-0189
17. McKinsey and Company, and The Business of Fashion (2024) The state of fashion 2024. McKinsey and Company. Accessed from https://www.mckinsey.com/. Accessed 6 Oct 2024
18. Global Brands Magazine (2024) AI's fashion revolution: Transforming the industry in 2024. Global Brands Magazine. Accessed from https://www.globalbrandsmagazine.com/ais-impact-on-fashion-industry-2024/. Accessed 6 Oct 2024
19. Ellen MacArthur Foundation (2021) Clothing doesn't need to be physical to exist: the Fabricant. Accessed from https://www.ellenmacarthurfoundation.org/circular-examples/the-fabricant/. Accessed 30 June 2025
20. European Commission, Joint Research Centre (2021) Textiles and the environment. Accessed from https://publications.jrc.ec.europa.eu/repository/handle/JRC139405_EN/. Accessed 30 June 2025
21. European Commission, Joint Research Centre (2024) Feasibility study of measures to incentivise improved textile waste management. Accessed from https://publications.jrc.ec.europa.eu/repository/handle/JRC139566/. Accessed 30 June 2025
22. European Environment Agency (2022) Management of used and waste textiles in Europe's circular economy. Accessed from https://www.eea.europa.eu/publications/management-of-used-and-waste-textiles/. Accessed 30 June 2025
23. Interreg North-West Europe (2022) Scaling textile recycling in Europe—turning waste into value. Accessed from https://vb.nweurope.eu/media/19019/2207-scaling-textile-recycling-in-europe-turning-waste-into-value.pdf/. Accessed 30 June 2025
24. Weinswig D, Bhatia M (2023) Digital merchandising strategies: the power of automation and 3D product imagery in retail. Coresight Research. Accessed from https://coresight.com/research/digital-merchandising-strategies-the-power-of-automation-and-3d-product-imagery-in-retail/. Accessed 30 June 2025
25. Business of Fashion and McKinsey and Company (2025) The State of Fashion 2025. Accessed from https://www.mckinsey.com/industries/retail/our-insights/state-of-fashion/. Accessed 30 June 2025
26. Shoham Y, Knight B, Hendrikson K (2025) 2025 AI Index Report. Human-Centered AI Initiative at Stanford University. Accessed from https://hai.stanford.edu/ai-index/2025-ai-index-report/. Accessed 30 June 2025
27. Adidas (2023) Annual Report 2023. Accessed from https://www.adidas-group.com, Accessed 10 Oct 2024
28. McKinsey and Company, and Global Fashion Agenda (2020) Fashion on climate: How the fashion industry can urgently act to reduce its greenhouse gas emissions. Global Fashion Agenda. Accessed from https://www.mckinsey.com/. Accessed 6 Oct 2024
29. Burberry (2023) Responsibility data appendix 2023/2024. Burberry Group plc. Accessed from https://www.burberryplc.com/. Accessed 6 Oct 2024
30. Farfetch (2023) Third annual Conscious Luxury Trends Report. Farfetch Limited. Accessed from https://aboutfarfetch.com/. Accessed 6 Oct 2024
31. Sustainability Magazine (2023) Top 10 brands embracing the circular economy. Sustainability Magazine. Accessed from https://sustainabilitymag.com/top10/top-10-brands-embracing-the-circular-economy-in-2023. Accessed 6 Oct 2024
32. Patagonia (2023) Patagonia Group Sales Fall 2023 Report. Patagonia, Inc. Accessed from https://www.patagonia.com/. Accessed 6 Oct 2024
33. Stella McCartney (2023) Impact Report 2023. Stella McCartney Ltd. Accessed from https://www.stellamccartney.com/. Accessed 6 Oct 2024
34. H&M Group (2024) H&M Group Annual and Sustainability Report 2023. H&M Group. Accessed from https://hmgroup.com/wp-content/uploads/2024/03/HM-Group-Annual-and-Sustainability-Report-2023.pdf/. Accessed 30 June 2025

35. The Sustainable Fashion Forum (2024) H&M is being sued for "misleading" sustainability marketing. What does this mean for the future of greenwashing? Accessed from https://www.thesustainablefashionforum.org/hm-greenwashing-lawsuit/. Accessed 30 June 2025
36. Modaes (2024) Adidas acusada de greenwashing: Una sentencia le obliga a retirar sus claims climáticos. Accessed from https://www.modaes.com/empresas/adidas-acusada-de-greenwashing-una-sentencia-le-obliga-a-retirar-sus-claims-climaticos/. Accessed 30 June 2025
37. Economic Times Brand Equity (2024) Why Burberry burns millions of dollars of clothes each year. Accessed 30 June–4 July 2025, from https://brandequity.economictimes.indiatimes.com/news/business-of-brands/why-burberry-burns-millions-of-dollars-of-clothes-each-year/115241715. Accessed 30 June 2025

Fashion Second-Hand Market and Re-Branding of Humana Portugal: One Strategy for Sustainability and Circular Economy

Júlia Queiroz, Teresa Raquel Barata, and Madalena Pereira

Abstract The transformation of the fashion ecosystem from a linear to a circular economy model is driving the development of innovative business models. One way to contribute to the circular economy is by extending the lifecycle of products, either through the purchase of higher-quality items or by enabling these items to have a second life through resale in second-hand clothing stores. This business model, already established in the market, particularly resonates with younger generations, such as Generation Z, both in online and offline retail environments. However, how can the value of second-hand fashion products and brands be effectively enhanced? One approach lies in the implementation of a branding strategy aligned with the values the brand seeks to communicate, supported by a coherent visual identity. This project aims to develop a re-branding strategy for a prominent second-hand fashion brand with national and international relevance: Humana. Based on established branding principles, the objective is to reposition the brand in a way that appeals more strongly to Generation Z consumers—thereby investing in the future while simultaneously fostering awareness of circular economy practices.

Keywords Re-Branding · Humana Portugal · Second-hand fashion stores · Extending lifecycle

J. Queiroz
Branding and Fashion Design, IADE-UE / UBI, Lisbon, Portugal

T. Raquel Barata · M. Pereira (✉)
FibEnTech, University of Beira Interior (UBI), Covilha, Portugal
e-mail: madalena.pereira@ubi.pt

M. Pereira
UNIDCOM/IADE-UE, University Europeia, Lisbon, Portugal

D. Raposo et al. (eds.), *Strategic Design for Sustainable Futures*, Springer Series in Design and Innovation 65, https://doi.org/10.1007/978-3-032-33282-0_2

1 Introduction

Considering the growing concerns associated with the fast fashion industry, explicitly those related to the excessive consumption of resources, excess production, which leads to the disposal of large quantities of clothes, and the emission of greenhouse gases, the fashion market for second-hand clothing has been growing in recent years. Although buying second-hand clothing is not a new practice, it has become a popular and more common in recent years, especially among younger individuals, due to the emergence of online stores that promote and develop their activity through the purchase and sale of second-hand clothing [1].

The traditional linear production model, currently still predominant in the fashion industry, particularly in fast fashion, based on the phases of design, production, use and disposal, results in the waste of large quantities of raw materials, which could be reused, and in the excessive consumption of natural resources, such as water and energy. The fashion industry continues to grow, largely due to the growth of fast fashion, which is based on low-cost production, rapid turnover of consumption and the short lifespan of items that end up being discarded and, for the most part, incinerated. The linear model promotes a rapid flow of materials, with fast cycles of design, production, consumption and disposal with numerous negative environmental and societal impacts. This model contributes to the short lifespan of products, with around 80% being discarded in just six months, and increases the business model of the fashion industry driven by increasing profit-seeking behaviors [2]. In the Linear Economy system, the main measure of value is the profit generated by a product in relation to its production cost, leading to an unsustainable production and consumption cycle characterized by high sales volume and low costs [3].

As sustainability becomes an increasingly central and urgent issue, brands must strive to offer products that are aligned with consumers' growing environmental concerns, through the implementation of supply chains and the adoption of production and operational practices aligned with the circular economy and sustainability [1].

The circular economy is a production and consumption model that aims to minimize waste or residues and maximize the use of resources, promoting the reuse, recycling and recovery of materials and products throughout their life cycle. By adopting the circular economy model, the fashion industry can reduce its environmental impact by promoting a closed product life cycle, with the aim of extending the useful life of garments. The circular economic approach makes it possible to maintain the value of products and materials for as long as possible [2].

In Portugal, Humana Portugal is an example of the commitment to sustainability and the promotion of the circular economy. Since its founding in 1998, this non-profit organization has been dedicated to preserving the environment through the reuse of textiles. In 2019, Humana published a report that showed that 30% of Portuguese consumers were willing to pay a higher price for natural, organic and environmentally sustainable products.

However, not all consumers are able to cover these additional costs, which makes Humana a key player in the supply of second-hand clothing. With 14 stores distributed between Lisbon and Porto, the organization guarantees access to second-hand clothing at affordable prices for a large number of people.

At the same time, the 2021 resale report from the ThredUp [4] platform points to significant growth in the second-hand clothing market in the coming years. The forecast indicates that this market could double in size in five years, with a particularly pronounced boost following the COVID-19 pandemic. According to the same report, by 2025, the market for second-hand clothing is expected to grow at a rate of 11 times greater than traditional clothing retail. This increase is attributed to the greater availability of carefully selected, high quality parts offered by sellers. In 2020, around 33 million consumers made their first second-hand clothing purchase, and 76% of them expressed their intention to continue shopping in this segment [4]. This change in behavior is strongly associated with generation Z—the group of individuals born approximately between the mid-1990s and early 2010s—who have been adopting more sustainable and circular consumption practices. The report also points out that, compared to the baby boomer generation, generally defined as those individuals born between the years 1946 and 1964, generation Z consumers are 165% more likely to consider the resale value of clothing when making a purchase and 83% more likely to have already sold items from their wardrobe [4].

For Gen Z, thrifting is more than just a way to shop, it's a lifestyle. Gen Z consumers want to be independent, save the planet, and save money, and thrifting allows them to wear a nice outfit that costs less than 10 dollars [5]. The main motivations for accessing clothing items through renting or purchasing second-hand clothing include also adopting more sustainable consumption habits by extending the life cycle of products.

The resale report from the ThredUp [4] highlights that the resale market is increasingly recognized as a promising growth channel in the retail sector. The results of a survey indicate that 62% of executives in this sector believe that their customers are already engaged in the resale of products, or that they have led many brands to adopt this trend in response to consumer demand. The report also reveals that 43% of consumers are more interested in purchasing from brands that offer credit in exchange for used items, while 34% prefer brands that offer both new and second-hand clothing. Additionally, 32% of consumers are more likely to associate a brand with high quality when it offers both types of products for purchase. Given this data, the importance of brand principles and values in building awareness and strengthening relationships with consumers becomes evident. The integration of resale into business strategies therefore emerges as an effective way of responding to public expectations and strengthening the connection between brand and customer.

According to Kotler [6] branding can be described as:

> To endow goods and services with the power of a brand, it's about creating mental structures and helping the consumer to organize their knowledge about the products in order to make your decision making clearer and, in this process, generate value for the company. [6].

The process of building a brand image can be understood as a reflection of its core business strategy. The establishment and evolution of a brand are considered fundamental mechanisms to promote organizational growth, allowing the introduction of complementary products or services, while increasing the visibility and positive perception of consumers in relation to the brand, particularly those linked to quality [7].

Based on the information presented above, this project intends to develop and implement a re-branding strategy for Humana Portugal, based on consolidated marketing and branding theories. The relevance of these concepts in promoting growth in the fashion resale sector is therefore recognized. Furthermore, the project aims to serve as a reference for other fashion brands with growth potential that, in the future, may face similar challenges and decide to consider re-branding processes.

1.1 Circular Economy and Sustainability

It is necessary to promote a transformation in the business model of the fashion industry, which must be based on the integration of sustainable practices throughout the supply chain and on consumer re-education, promoting more conscious and responsible purchasing behaviours. Sustainability and circularity are fundamental concepts in this transformation, aiming to reduce the environmental and social impacts of the fashion industry. Sustainability has been defined as the "development that meets the needs of the present without compromising the ability of future generations to meet their own needs" [8].

Sillanpää and Ncibi [9] state that the principles underlying the Circular Economy are not recent, but date back to the earliest communities that lived in harmony with the cycles of nature. In these early communities, resource use was inherently sustainable, as they were closely dependent on natural systems. Over time, however, the relationship between human activities and the environment has changed, particularly with the industrial revolution and demographic growth. The industrial, agricultural and technological revolutions changed the relationship with the environment, moving towards extensive extraction and control of natural resources. The development of tools and processes has made it possible to manipulate and exploit natural resources in an increasingly effective and systematic way. The fundamental concepts of the Circular Economy were further developed in 1976 by Stahel and Ready [10], who, in the context of the industrial economy, proposed an economic model structured as a closed loop. They suggested strategies aimed at minimizing waste production, achieving efficiency in the use of resources, encouraging dematerialization and creating employment opportunities through innovative industrial practices [10]. The current Circular Economy model seeks to reduce the use of natural resources by promoting a restoration and regeneration approach, closing material cycles, reducing waste and transforming the economic models towards greater environmental responsibility. The circular economy model has as its main objectives fewer raw materials, less waste and fewer emissions. The life cycle of products is extended through the reuse, repair, refurbishment and recycling of materials and products as much as possible. Globalization as a business model of the current Industrial Economy is not viable in the long term. Industrialized countries must implement new business models that will enable them to make the transition to sustainable development. These new models include clean production processes and eco-design of goods, replacing end-of-life with a cradle-to-grave cycle to close the loop and reduce waste [11]. In the context of sustainability and lifecycle analysis it is used to encompass the entire lifespan of a product, from its initial raw material source to its final disposal or recycling. Extending the use of clothing items, i.e., extending their life cycle, allows us to reduce consumption and mitigate environmental impacts [12]. Eco-design principles, such as design for recyclability, modular design, multifunctionality, biodegradability and design for disassembly, must be incorporated into product development from the early stages of design [13].

According to Kjaerheim [14], cleaner production (CP) involves using all necessary resources to effectively eliminate toxic raw materials and reduce the amount of toxicity from all emissions and waste. CP has been shown to be an effective way to achieve better and more efficient use of materials, lower energy consumption and a reduction in the production of greenhouse gases, a reduction in emissions to air, water and soil, a reduction in the use of toxic materials in production and products and a reduction in the amount of waste in landfills [14]. CP model also promotes positive preventive actions and a holistic view of resources, production, economy, quality, environment, labour relations, human resources, finite/renewable materials and energy resources and profit generation [14].

Increasing consumer knowledge and awareness is crucial to promoting sustainable consumption behaviours. Therefore, it is essential to provide adequate and comprehensive information about the sustainability of clothing—covering product attributes prior to purchase, maintenance practices and lifespan extension during use, and reuse or recycling services after use [12]. In this context, the Circular Economy can be defined as:

> Circular Economy as a regenerative system in which resource input and waste, emission, and energy leakage are minimized by slowing, closing, and narrowing material and energy loops. This can be achieved through long-lasting design, maintenance, repair, reuse, remanufacturing, refurbishing, and recycling. ([15], p. 7).

The Circular Economy is proposed as a model based on three main pillars: the elimination of waste and emissions, the use of products for as long as possible and the regeneration of natural systems. This approach contributes significantly to addressing global challenges, while offering benefits to companies in economic terms, people in social terms, and the environment [16]. It involves a wide range of sustainable practices, such as repair, reuse, recycling, eco-design and service-based models, all of which are subject to constant evolution. According to the publication on the theme of the new textile economy by the Ellen MacArthur Foundation [17], the Circular Economy represents an alternative to the classic linear model of "take-make-dispose," by presenting as objectives the reduction of waste generation and its environmental impacts, encouraging the use of biodegradable materials whenever possible and promoting the reuse, replacement and recycling of non-biodegradable elements. In this context, product design must be directed to facilitate the integration of products into closed life cycles, allowing the continuous reuse of materials and components.

Therefore, clothing should be designed to allow its use for as long as possible, facilitating its repair, reuse and upgrade, as well as its remanufacturing and recycling, at the end of use, rather than being discarded. Additionally, production processes must consider the reuse of materials and components, as well as the regeneration of natural resources.

Figure 1 shows the lifecycle of second-hand textiles. A range of 38% textiles are separately collected and sent to charity shops and bring bank for reuse and recycling. According to the conditions, textiles are sorted and roughly 10% high quality textiles are resold in Europe. The remainder is exported to other countries, namely Asia or Africa.

Within the circular economy, the cycle is completed with the conversion of waste into resources. According to Coltelli and Morganti [3], the Circular Economy must operate based on the following ten key principles that collectively promote sustainability, resource efficiency, and environmental and social responsibility. The first is the concept of waste as a resource, which involves the conversion of waste into new materials, namely non-biodegradable waste. The second principle is use, which refers to the reintroduction into the economic circuit of products that no longer meet the initial needs of consumers. The third principle is reuse, which involves using products or their components to make new products. Next comes repair, which aims to extend the useful life of damaged products, and recycling, which focuses on recovering and using the materials present in the waste. The sixth principle is valorisation, which involves deriving energy from waste that cannot be recycled, and the seven is the use of energy from renewable sources in the production, reuse and recycling processes, prioritizing renewable energy. Eco-design constitutes the eighth principle that refers to the consideration and integration of environmental impacts throughout the entire life cycle of a product. The ninth is industrial and territorial ecology, which aims at the optimized management of stocks and flows of materials, energy and services. Finally, the tenth principle is the functional economy, which involves the implementation of new business models.

1.2 Branding

According to Keller and Lehmann [19], brands basically "serve as markers for the offering of a firm." Building a brand begins with the product itself and is subsequently shaped by marketing actions and the use that customers make of the product, which is influenced by their experience with it [19]. Furthermore, the same authors state that "brands manifest their impact at three main levels: customer market, product market and financial market."

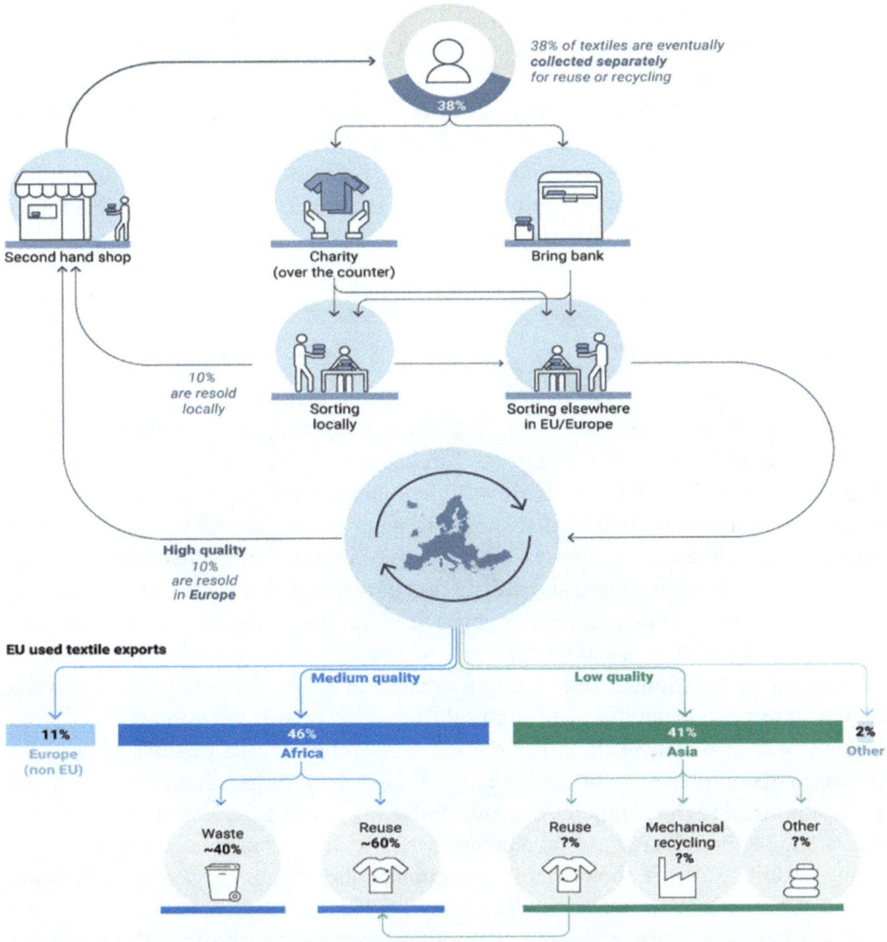

Fig. 1 Circulation of worn-out or second-hand textiles collected across European countries. *Source* [18]

Branding consists of the process of creating and managing the identity, image and perception of a brand among consumers. To establish an active presence in the mind of consumers, it is necessary integrate various intangible elements associated with the brand, such as awareness, attributes, benefits, images, thoughts, feelings, attitudes and experiences. The brand creates a personal and social identity, to promote a feeling of belonging or distinction in consumers and is recognized through signs and symbols, which can be perceived positively or negatively. Branding consists of shaping the way a brand is recognized, remembered and emotionally connected to the customer, through visual elements such as the name, logo and design, and through the message transmitted and the overall experience it provides. The visual elements are usually applied directly to a product or indirectly through labels, serving as markers that can positively distinguish a brand [20]. At its essence, branding is not just about visual elements but about building a consistent and lasting relationship between the brand and its customers.

As a business practice, branding is often described as a promise made by an organization to its target customers, which reflects not only functional benefits but also emotional, self-expressive and social values. Furthermore, brands function as fundamental tools in their relationships with customers and influence both their market positioning and their financial performance [7].

1.3 Branding Identity

As consumers have an increasing number of choices, companies must use creativity to create compelling brand experiences that stick in the minds of consumers to attract new customers and increase loyalty among existing ones. Furthermore, every customer interaction with the brand should be seen as an opportunity to deepen their engagement with it. In this context, the concept of brand identity also plays a fundamental role in the construction and affirmation of a brand. Brand identity is a tangible element that uses a human-centered approach to increase the differentiation of organizations, add value, make relevant ideas accessible and include all these aspects in a unified system should use multiple communication channels to reach diverse audiences to raise awareness and highlight the strengths of the company or brand, allowing it to position itself in the market and emphasize its positive differences compared to companies in the same sector. A successful brand identity should focus on communicating the company's unique and distinctive value, leading customers to understand the brand's features and benefits, and fostering customer recognition and loyalty. Additionally, it should include elements such as an easy-to-identify name and a distinctive and attractive product design [21].

When building a brand identity, there are other elements besides the name that are essential, such as logos, symbols and slogans. Furthermore, various concepts are used to build brand identity, such as significance, memorability, aesthetic appeal, adaptability, and long-term flexibility, and these concepts may evolve during the development of the brand's verbal and visual identity [19]. As Wheeler and Meyerson [21] state:

> Brand awareness and recognition are facilitated by a visual identity that is easy to remember and immediately recognizable. Visual identity triggers perceptions and unlocks associations of the brand. Sight, more than any other sense, provides information about the world. ([21], p. 20)

For Aaker [7], brand identity is also understood as a brand vision, a systematic structure to distinguish the brand from market alternatives. It is a description of the intended identity for the brand and the perception that the company wants to create among customers, employees and partners. Brand vision is a fundamental component in terms of developing a company's marketing programs and strategic planning.

Refreshing the brand vision, when executed effectively, can redefine the business strategy of the company, differentiate the brand from competitors, increase customer engagement, inspire employees, and generate new ideas for marketing campaigns. According to Aaker [7], the process of developing a brand vision should follow the next three steps: in the first step, an exhaustive analysis of customer segments, competitors, market insights, influence and weaknesses of the brand, as well as the business strategy that will support it, should be carried out; in the second stage, all the intended associations must be defined to guide the brand's value proposition; In the third stage, the elements that make up the brand and its main characteristics must be identified, which will help in designing the brand's vision [7].

2 Methodology

2.1 Objectives

The aim of this project was to increase the visibility of the non-profit organization Humana in Portugal, which has been promoting textile reuse since 1998, positioning it as a leading player in the national second-hand clothing market. To achieve this, the project focused on rebranding the company to create a more attractive image that would resonate with Generation Z customers, who are more likely to engage with the second-hand market based on the company's values. With this in mind, the project aimed to strengthen brand awareness by developing a strong brand identity that is recognizable and memorable for customers, through the creation of a distinctive and cohesive visual identity that included the brand's DNA, image and identity. During the project, a new logo, brand labels and social media content were developed in line with the new brand image. In addition, the rebranding will update the brand's website, marketing materials and in-store signage to reflect the new brand identity.

2.2 Project Methodology

The design methodology adopted in this work followed the brand DNA process, using the Brand DNA Tool® [22–24], with the objective of developing a rebranding strategy for the Humana Portugal brand—a second-hand fashion brand with national and international relevance. This methodology was chosen based on its effectiveness and the proven positive results it has demonstrated in the areas of branding and rebranding. The brand DNA process is a comprehensive and systematic framework that allows defining and articulating the primary elements of a brand identity [25]. It allows, following a series of steps, the development of a clear and appealing brand identity that can be used in all media and moments of contact with the public [26]. To create a Brand's visual identity, Oliveira [24, 27], Olins [28], Raposo [29], Lupton and

Phillips [30] and Wheeler and Meyerson [21], state that the visual identity elements are the graphic brand, typography, symbols, colors, graphics, name, imagery, form and movement, sound, and the so-called "fifth element." These elements must be, both graphically and semantically, aligned with the brand's DNA, so that there is a consistent relationship and a degree of visual coherence that facilitates recognition. The project was structured as follows: research, competitive analysis, development of the Humana Portugal persona, SWOT analysis, application of the Brand DNA process tool to the Humana brand, brand positioning, inspiration research, logo and design exploration, promotional launch strategy, and final project delivery. The rebranding project aimed to increase brand recognition, promote customer loyalty and attract a new generation of customers who value sustainability and ethical practices.

3 Analysis and Results

3.1 Analysis of Brand Humans Portugal

To stand out from its competitors, Humana Portugal undertook a brand DNA process to identify the unique qualities, characteristics, and attributes that define its core principles. This process focused on four key elements essential to understanding the brand's distinct identity: Vision, Mission, Values, and Personality. Humana Portugal's mission is to promote a circular economy by giving used textiles a second life, embodying a sustainable, second-hand approach. Based in Alcochete, Lisbon, Humana Portugal is a non-profit organization committed to environmental and social responsibility.

Humana Portugal is part of the Humana People to People Federation, a global network of 29 non-profit organizations dedicated to international solidarity, development, and cooperation [31]. The organization promotes textile reuse by collecting donations through containers placed in Lisbon, Porto, and across the country, operating in a total of 879 collection points from north to south. These donations are processed and sold in second-hand stores, supporting circular economic efforts and reducing textile waste.

In 2022, Humana collected 2,777 tons of used clothing, preventing them from ending up in landfills or being incinerated. According to the Humana People to People Federation, reusing one kilogram of clothing avoids the emission of 6,1 kg of CO_2, meaning Humana's efforts prevented 16,939 tons of CO_2 emissions that year [31]. As a pioneer in selective textile collection and sustainable waste management, the organization transforms surplus clothing into valuable resources with social impact.

By participating in Humana's initiatives, partner organizations benefit from a sustainable solution for managing excess clothing and footwear, improved corporate social responsibility, employee involvement, and the chance to support meaningful social programs.

When examining Humana's customer personas for this rebranding initiative, it can be divided into three main distinct groups based on an individual's shopping habits. The first is Mariana, a Gen Z shopper who values mindful consumption and supports brands aligned with slow and sustainable fashion principles. The second is Inês, a baby boomer who shops at Humana for its affordability and the convenience of buying for both herself and her family. The third is André, a millennial drawn to thrift shopping for the thrill of finding unique vintage pieces. These personas all differ greatly, however, they reflect Humana's broad customer base, which includes socially conscious Gen Z buyers, cost-aware boomers, and millennials seeking nostalgic fashion. By recognizing and addressing the unique preferences of each group and the foundation of the brand, Humana can strengthen its customer engagement and refine its brand identity through the rebranding effort.

In order to gain a deeper understanding of Humana, a SWOT analysis was conducted to identify and evaluate the main internal and external factors that could influence the organization's success. This analysis helps guide the development of a strategic plan by leveraging strengths and opportunities while addressing weaknesses and potential threats (Fig. 2).

Humana is becoming a leading organization in the collection and resale of textile materials, using its profits to fund social initiatives that improve key areas such as education, healthcare, agriculture, and clean energy in Africa. Its mission focuses on making textile donations easy and accessible by providing collection containers across the country. In addition, Humana Portugal collaborates with businesses and government entities to collect surplus clothing and textiles from the fashion industry. The brand's core values emphasize both social and environmental sustainability,

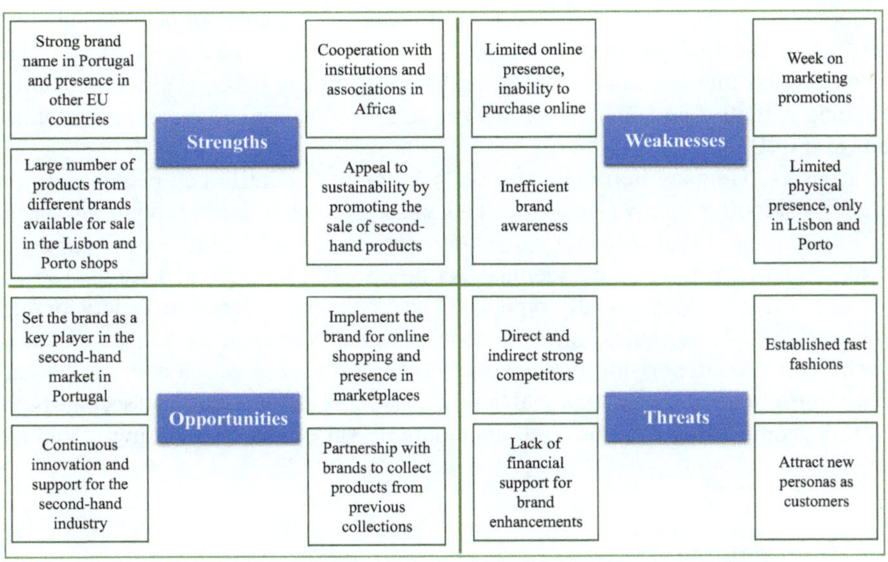

Fig. 2 SWOT analysis. *Source* Authors

Fig. 3 Typography, colors, logo and secondary elements of Humana Portugal. *Source* Humana Portugal

with a primary focus on reselling quality second-hand clothing to promote a circular economy. These efforts are complemented by reinvesting profits into international social development projects. Humana's brand personality is defined as fair and informative, appealing to both men and women looking for affordable, high-quality fashion.

The analysis of Humana Portugal's visual identity highlights key design elements such as typography, color, and graphic branding. The typography used is from the Akagi family, including variations like Akagi Pro, Bold, Light, and Regular. The color palette consists primarily of green and white. According to Nielsen [34], these colors in web design often convey values such as eco-consciousness, purity, and environmental responsibility. The graphic mark features the logo of People to People International (PTPI), a non-profit organization to which Humana belongs. This logo, which includes a globe, reflects PTPI's mission of fostering peace and mutual understanding across cultures through educational and cultural exchanges. These secondary visual elements play an essential role in strengthening Humana Portugal's brand identity, aligning with its core values and contributing to a cohesive and recognizable brand image (Fig. 3).

To assess Humana Portugal's market position and identify competitive gaps, a brand positioning map was created. This strategic tool helped pinpoint key areas of differentiation that resonate with consumers and can be used when shaping a distinct and appealing brand identity. As described by Kotler and Keller [26], a brand positioning map visually represents how consumers perceive various brands based on specific attributes, allowing companies to understand their relative market position. It also supports the creation of a unique value proposition that sets a brand apart from competitors. For this analysis, both national and international second-hand clothing competitors were selected, including depop[1]; Vinted[2]; Flamingos Vintage

[1] https://www.depop.com/.

[2] https://www.vinted.pt/.

Fig. 4 Analysis of the competitive brands in the market and positioning of the Humana brand

kiko Lisboa,[3] The RealReal,[4] Vestiaire Collective,[5] Dona Ajuda,[6] ThreadUp[7] and A Outra Face da Lua.[8] The findings indicate that Humana is currently viewed as a low-cost and basic option, lacking emotional resonance with consumers. However, there is a considerable opportunity to reposition the brand as both stylish and affordable (Fig. 4).

Humana Portugal's current branding is considered outdated and poorly aligned with the fashion market. Its design lacks modern appeal and has not evolved with shifting consumer trends. Additionally, the absence of an online store and presence on platforms like TikTok limits its reach and responsiveness to customer expectations. To strengthen brand recognition and reposition Humana as both fashionable and affordable, this project proposes a rebranding strategy aimed at attracting a wider audience, particularly Gen Z and Millennials. The initiative will focus on building a cohesive brand image across all channels while prioritizing key digital touchpoints, including social media engagement and online reputation management.

According to Wheeler and Meyerson [21], Oliveira [24], and Raposo et al. [32, 33], brand language reflects the brand's overall mood and is composed of visual elements and graphic symbols that express its identity and personality. After conducting a competitor analysis and examining Humana's brand DNA, mood boards and logo representation boards were developed to help define the brand's visual style (see Fig. 5 for an example from Vestiaire Collective).

Staying true to Humana's core values of social and environmental sustainability, the newly developed logo reflects this vision. The brand retained green as its primary

[3] https://flamingoslisboa.com/.

[4] https://www.therealreal.com/.

[5] https://www.vestiairecollective.com/.

[6] https://donaajuda.pt/.

[7] https://www.thredup.com/.

[8] https://aoutrafacedalua.com/.

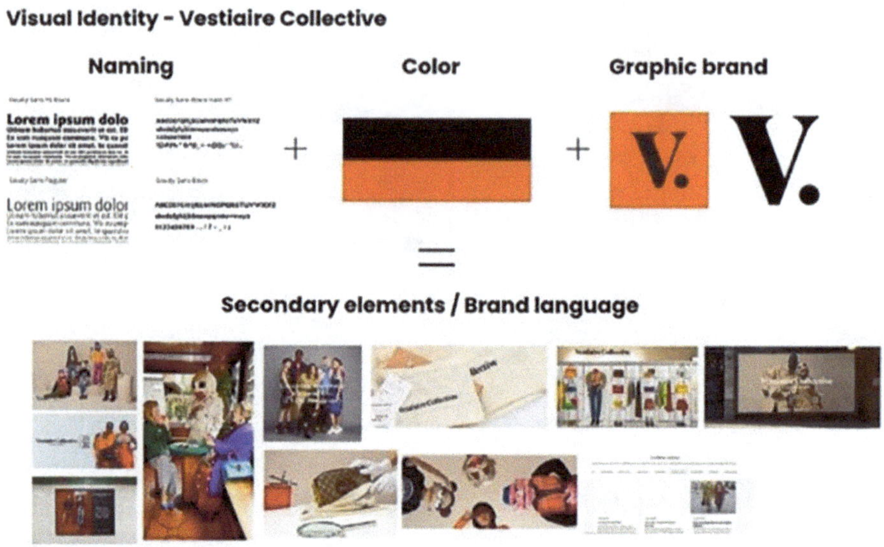

Fig. 5 Example of visual identity analysis

color to symbolize eco-friendliness and purity, but adopted a brighter, gradient variation to modernize its appearance. The logo also incorporates different representations of the letter "H," symbolizing the brand's name.

Three serif typefaces were selected to support the visual identity. Beatrice Regular offers sharp, angular serifs with geometric forms, giving it a contemporary edge. Beatrix Antiqua, with its soft curves and classic style, adds versatility for applications like branding, editorial, and packaging. For digital communications and social media, the sans-serif Poppins typeface was chosen for its clean, modern look suitable for web and branding use.

Two main logo variations were created, one with a green background and a white "H," and another with a white background and a green "H." Additional versions were also designed using the brand's green-and-white color palette. These visual assets align with Humana's mission and identity while providing flexibility for diverse design applications (see Figs. 6, 7, 8, 9 and 10).

4 Conclusions

As consumers become more aware and concerned about the environmental and social impact of the fast fashion industry, they are also increasingly looking for more ethical and sustainable alternatives that offer exclusivity at the same time. This change in consumer attitudes, coupled with recent regulations, has also encouraged the fashion

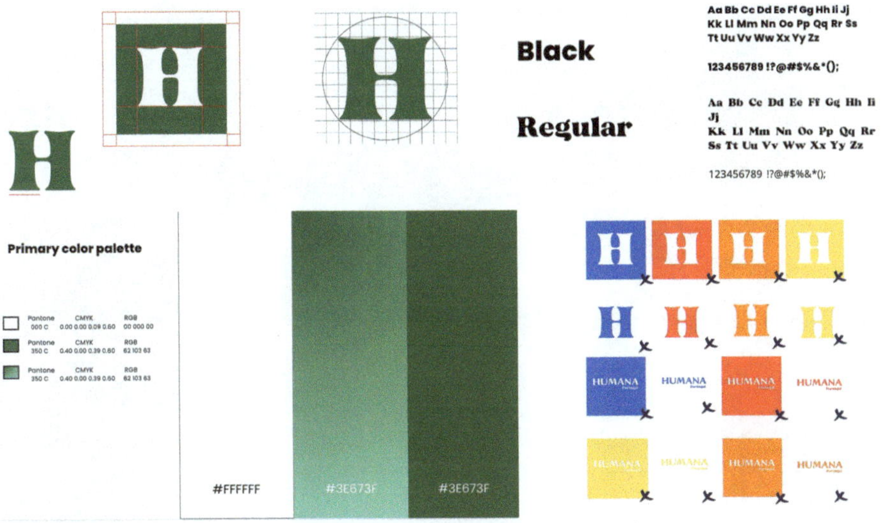

Fig. 6 Logo construction, Beatrice typography, color palette with logo variations not allowed. *Source* Authors

Fig. 7 Visual identity and eco-bags. *Source* Authors

industry to change traditional linear production models to adopt production and operating practices in line with the circular economy and sustainability.

The growing popularity of the second-hand fashion market is due to the increasing concerns about the environmental impact of fast fashion and the desire for more sustainable and unique fashion options. This trend is particularly evident among Generation Z and Millenniums consumers, who are more likely to buy and sell second-hand clothing than other generations. This new generation of consumers is very conscious of their impact on the environment and seek out ethical and sustainable fashion options but still want to wear trendy second-hand clothing.

This rebranding project of the Humana Portugal brand has shown that it can become the first major player in the second-hand clothing market in Portugal, since, according to the competitive analysis, there are not many brands in Portugal with the

Fig. 8 Bags in recycling paper and tags with logos. *Source* Authors

same characteristics or potential. The rebranding, which aims to increase awareness, promote customer loyalty and attract a new generation of customers who value sustainability and ethical practices, will allow Humana to expand its presence throughout the country.

In addition, during this study it was also found that a visual merchandising project should be applied to put into practice all the rebranding techniques developed for the company, as well as to increase sales and attract new customers. Since visual merchandising design requires the creation of strategies that combine creativity, design skills and marketing techniques, it was not possible to cover all these aspects in this article.

In conclusion, this study highlights the importance of sustainability and ethical practices in the fashion industry, especially among Generation Z. It also emphasizes the importance of rebranding as a strategy for building brand awareness and attracting new customers. Overall, the ideas presented in this article can serve as a valuable reference for real-life market scenarios and can stimulate academic discussions about sustainability in the fashion industry and rebranding projects.

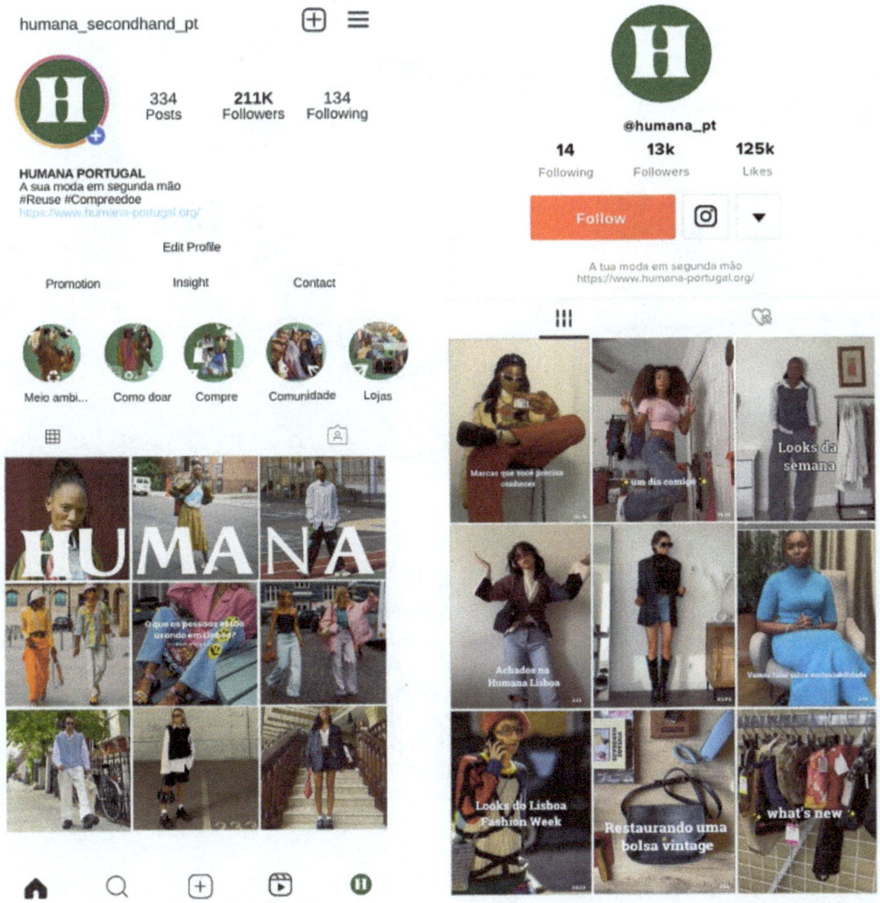

Fig. 9 Instagram and TikTok examples. *Source* Authors

Fig. 10 Outdoor example. *Source* Authors

Acknowledgements The authors acknowledge the financial support from integrated project be@t—Textile Bioeconomy (TC-C12-i01, Sustainable Bioeconomy No. 02/C12-i01.01/2022), promoted by the Recovery and Resilience Plan (RRP), Next Generation EU, for the period 2021–2026.

The authors are very grateful for the support granted by the Research Unit of Fiber Materials and Environmental Technologies (FibEnTech-UBI), through the Project reference UIDB/00195/2020, funded by the Fundação para a Ciência e a Tecnologia, IP/MCTES through national funds (PIDDAC), and https://doi.org/10.54499/UIDB/00195/2020.

References

1. McKinsey and Company: How online marketplaces are making secondhand fashion a first choice (2021). https://www.mckinsey.com/industries/technology-media-and-telecommunications/our-insights/how-online-marketplaces-are-making-secondhand-fashion-a-first-choice. Accessed 21 Dec 2021
2. Niinimäki K (2018) Sustainable fashion in a circular economy. Aalto University, Helsinki
3. Coltelli MB, Morganti P (2020) An introduction to the circular economy. Springer, Cham
4. TthredUP (2021) Resale Report: Resale Industry. https://www.thredup.com/resale/2021/#resale-industry. Accessed 5 Apr 2025
5. Huber E (2020) An investigation into why thrifting is such a defining aspect of Gen Z life. https://www.refinery29.com/en-us/2020/10/10014753/thrifting-gen-z-thrift-shopping-trend. Accessed 17 Dec 2023
6. Kotler P (2012) Administração de marketing. Prentice Hall, São Paulo
7. Aaker D (2014) Aaker on branding: 20 principles that drive success. Morgan James Publishing, New York
8. Brundtland GH (1987) Our common future: report of the world commission on environment and development. Oxford University Press, Oxford
9. Sillanpää M, Ncibi C (2019) The circular economy: case studies about the transition from the linear economy. Academic, London

10. Stahel W, Reday G (1976) The potential for substituting manpower for energy: report to the commission of the european communities. European Commission, Brussels
11. Stahel WR (2010) The performance economy, 2nd edn. Palgrave Macmillan, London
12. Schiaroli V, Fraccascia L, Dangelico RM (2024) How can consumers behave sustainably in the fashion industry? a systematic literature review of determinants, drivers, and barriers across the consumption phases. J Clean Prod 483:144232
13. Casciani D, Wang W (2023) Unpacking strategies for E-textile design for disassembly. In: Heinzel T, Dumitrescu D, Tomico O, Robertson S (eds) Proceedings of textile intersections conference. London, United Kingdom
14. Kjaerheim G (2005) Cleaner production and sustainability. J Clean Prod 13(4):329–339
15. Geissdoerfer M, Savaget P, Bocken NMP, Hultink EJ (2017) The circular economy—a new sustainability paradigm? J Clean Prod 143:757–768
16. Ellen MacArthur Foundation: Fashion and the Circular Economy (2021). https://archive.ellenmacarthurfoundation.org/explore/fashion-and-the-circular-economy. Accessed 11 Nov 2024
17. Ellen MacArthur Foundation A New Textiles Economy: Redesigning Fashion's Future (2021). https://emf.thirdlight.com/link/2axvc7eob8zx-za4ule/@/preview/1?o. Accessed 11 Mar 2025
18. Kohler A, Watson D, Trzepacz S, Löw C, Liu R, Danneck J (2021) Circular economy perspectives in the eu textile sector. European Union, Brussels
19. Keller KL, Lehmann DR (2006) Brands and branding: research findings and future priorities. Mark Sci 25(6):740–759
20. Bastos W, Levy SJ (2012) A history of the concept of branding: practice and theory. J Hist Res Mark 4(3):347–368
21. Wheeler A, Meyerson R (2024) Designing brand identity: a comprehensive guide to the world of brands and branding, 5th edn. John Wiley & Sons, Hoboken
22. Olhats M (2012) Decoding the brand DNA: a design management methodology applied to favela fashion. Repositório UFSC, Florianópolis
23. Gomez LSR, Olhats M, Floriano J, Vieira MLH (2011) Fashion's brand DNA: O Processo. Vida Económica 145
24. Oliveira RM (2015) Design de Marcas: Construindo a Expressão da Identidade Corporativa. Bookman, Porto Alegre
25. Kapferer J-N (2012) The new strategic brand management: advanced insights and strategic thinking, 5th edn. Kogan Page, London
26. Kotler P, Keller KL (2016) Marketing management, 15th edn. Pearson, Upper Saddle River
27. Oliveira F (2018) Chapter seven: brand identity and visual culture. In: Raposo D (ed) Communicating visually: the graphic design of the brand. Cambridge Scholars Publishing, Newcastle upon Tyne, pp 117–130
28. Olins W (2014) Brand new: the shape of brands to come. Thames & Hudson, London
29. Raposo D (ed) (2018) Communicating visually: the graphic design of the brand. Cambridge Scholars Publishing, Newcastle upon Tyne
30. Lupton E, Phillips JC (2008) Graphic design: the new basics. Princeton Architectural Press, New York
31. Humana Portugal (2020) Juntos Somos Mais Fortes. https://www.humana-portugal.org/wp-content/uploads/2020/06/periodico-octubre-portugues-web.pdf. Accessed 5 Apr 2024
32. Raposo D, Oliveira F, Farinha L (2020) From identity into brand visual identity: finding and defining the intangible brand DNA. In: Handbook of research on driving industrial competitiveness with innovative design principles. IGI Global, Hershey, pp 66–85
33. Raposo D (2025) Design-driven branding: a proposal for a methodological model for strategic and creative brand direction and design of the brand visual identity. In: Brandão D, Martins N, Duarte E (eds) Perspectives on design and digital communication V. Springer series in design and innovation, vol 50. Springer, Cham, pp 157–172
34. Nielsen J (1995) Web design for the human eye. Nielsen Norman Group, Fremont

Generative Design and Additive Manufacturing: An Analytical Approach to Personalization in the Footwear Industry

Cláudia Liberato Nunes, Miguel Terroso, and Nuno Martins

Abstract Generative design has emerged as an innovative approach to footwear personalization, enabling highly optimized models to be developed that conform to the anatomical characteristics of the human foot. From this standpoint, the present study explores its impact, emphasizing its potential to address the limitations of mass production. Moreover, integrating generative design with additive manufacturing makes it possible to create complex, adaptive geometries that enhance structural efficiency and improve ergonomic fit. This research focuses on analysing lattice structures and evaluating their performance in terms of impact absorption, weight reduction, and structural support. Computational simulations and structural analyses are employed to optimize the product before manufacturing, providing accurate predictions of mechanical performance and ensuring its viability. Additionally, multiple software tools are compared, underscoring the significance of automation and simulation integration in the design process. The findings demonstrate that this methodology fosters the development of personalised, high-performance, and sustainable footwear solutions, establishing a new paradigm for the footwear industry. Specifically, this approach improves production efficiency and user comfort through digital innovation, while reducing material waste and allowing a more precise adaptation to individual biomechanical needs.

Keywords Generative design · Additive manufacturing · Footwear personalization · Structural optimization · Lattice structures

C. Liberato Nunes (✉) · M. Terroso · N. Martins
Media and Culture, School of Design, ID+ Research Institute in Design, Polytechnic Institute of Cavado and Ave, Barcelos, Portugal
e-mail: cliberato.nunes@gmail.com

M. Terroso
e-mail: mterroso@ipca.pt

N. Martins
e-mail: nmartins@ipca.pt

D. Raposo et al. (eds.), *Strategic Design for Sustainable Futures*, Springer Series in Design and Innovation 65, https://doi.org/10.1007/978-3-032-33282-0_3

1 Introduction

The footwear industry is significantly transforming, driven by environmental challenges, technological advancements, and ergonomic demands [1]. Traditionally, mass production has relied on standardised manufacturing processes, where footwear design is constrained by generic lasts that don't consider the anatomical diversity of human feet [2–4]. Consequently, personalization, comfort, and biomechanical performance are often compromised, leading to premature disposal, improper plantar pressure distribution, and an increased risk of injury [5, 6].

Simultaneously, growing environmental awareness and rapid technological progress have intensified consumer demand for products that reflect identity [7, 8]. This shift encouraged brands to adopt digital manufacturing strategies and made-to-measure personalization, offering solutions that align with market expectations [9].

In response to these challenges, this research explores generative design as a disruptive strategy within the footwear sector, enabling tailored, sustainable, and technologically advanced solutions. The research question is therefore: How can generative design improve footwear development? This reflects a broader effort to explore how digital tools can address key industry needs. The hypothesis is that integrating algorithmic design processes enables more personalized, mechanically efficient, and environmentally responsible objects. Compared to conventional methods, this data-driven approach aims to reduce material waste, improve ergonomic fit, and streamline product development.

The study adopts an exploratory and analytical methodology, combining literature review with a comparative evaluation of digital tools used in generative design. Special attention is given to the role of lattice structures and software integration in promoting efficient design and production workflows.

This document is structured into five sections to provide a comprehensive analysis: Sect. 2 discusses the challenges of the footwear production cycle; Sect. 3 reviews the theoretical and technical foundations of the project; Sect. 4 evaluates the software tools available for model development; and finally, Sect. 5 presents the main conclusions and future perspectives.

2 Industry Challenges Throughout the Product Life Cycle

The scale of footwear production and consumption has never been higher. The industry faces severe environmental challenges, with 34 million pairs sold daily and an average of four pairs per person annually. The prevailing mass production model, characterized by rapid manufacturing and disposal, significantly impacts the entire product life cycle, a situation further exacerbated by the rise of fast fashion [10, 11].

When analysing the industry's environmental footprint, studies estimate that each pair of shoes generates an average of 14 kg of CO_2, with the manufacturing phase being the most environmentally detrimental. Additionally, each shoe may contain around 40 different materials, including adhesives, plastics, and textiles, which are incorporated in ways that make reuse and recycling nearly impossible [10, 12].

The manufacturing stage involves approximately 360 procedures, combining industrial and artisanal techniques. Common operations include cutting, sewing, injection moulding, sole pressing, and final assembly, which require equipment including hydraulic presses and automated bonding systems. Furthermore, processes like dyeing and adhesive application consume large volumes of energy and water while emitting toxic substances that pose risks to workers' health and the environment [12, 13].

The fragmentation of the supply chain further amplifies this impact, as reliance on multiple suppliers and globally dispersed factories increases CO_2 emissions from the transportation of raw materials and semi-finished products [14].

At the end of its life cycle, 80% of discarded footwear is sent to landfills, releasing harmful chemicals into the soil and groundwater, while the remaining 20% is incinerated, contributing to air pollution and greenhouse gas emissions [11, 12].

Considering these challenges, this study adopts an analytical approach to explore how generative design and additive manufacturing can provide viable solutions and redefine the footwear industry paradigm.

3 State of Art

3.1 Generative Design Methodology

This research applies generative design techniques to develop optimized lattice structures for footwear. The approach uses algorithms to create adaptive geometries based on design constraints, such as ergonomic requirements, material performance, and production technologies. Similar to topology optimization, this method aims to minimize mass and maximize structural efficiency, key factors in footwear design [15].

The technological foundation of generative design incorporates neural networks and artificial intelligence, facilitating the automatic adaptation of designs to specific project requirements. This capability enables the exploration of innovative solutions that would be unattainable through conventional design methodologies [16]. As outlined in Fig. 1, the generative design process is structured into four primary stages:

1. Design Variables—Establishing design criteria such as overall geometry, mechanical behavior, and expected performance.
2. Generation—Processing input data to generate multiple design options through algorithms and constraints.

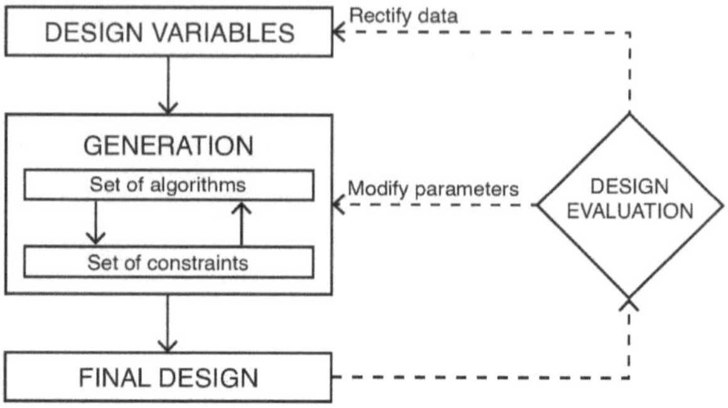

Fig. 1 Iterative approach of generative design

3. Final Design—Identifying the most effective solution based on defined input parameters, followed by computational validation through simulation techniques.
4. Design Evaluation—Iteratively refining the design by adjusting parameters and rectify data until a satisfactory solution is achieved [15, 17].

The implementation can be achieved through Visual Programming Language (VPL) or Text-Based Programming Language (TPL). VPL, used in platforms such as Grasshopper, Dynamo, and nTopology, enables the visual creation of designs by connecting graphical blocks representing mathematical operations and parametric relationships. This method is advantageous for designers and engineers who prefer an intuitive workflow that eliminates extensive coding expertise. In contrast, TPL is based on programming languages such as Python, C#, and MATLAB, offering greater flexibility and precision. It allows the development of highly complex scripts, enabling greater control over algorithmic processes and computational efficiency [18].

3.2 *Parameter Setting*

The definition of constraints and parameters directly impacts the quality of the solutions generated, as it establishes the guidelines that direct the algorithms during generative design [16]. In this context, the parameters illustrated in Fig. 2 were defined to inform the design process:

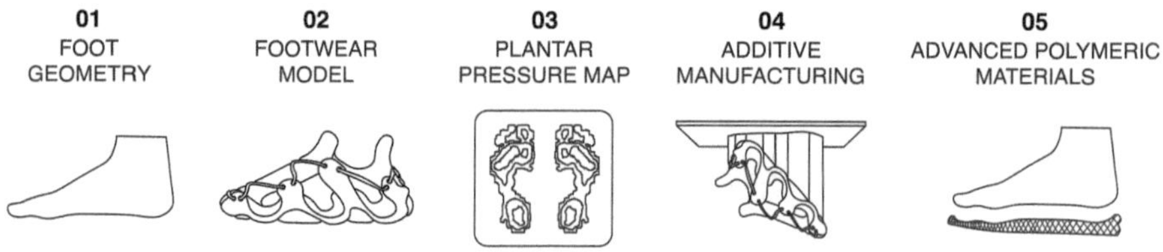

Fig. 2 Design parameters for personalized footwear development

1. Foot Geometry—A 3D-scanned model capturing the user's anatomy. This method obviates the need for standardized lasts, reducing discomfort and premature product disposal [6, 19, 20].
2. Footwear Model—A CAD-based framework that defines the footwear's shape and density. The design is adjusted according to the digitized morphology, ensuring an optimized internal structure tailored to each foot [21, 22].
3. Plantar Pressure Map—A graphical representation of load distribution across the foot, identifying higher and lower-pressure areas under static and dynamic conditions. This data helps optimize impact absorption, reinforcing high-stress areas and reducing not only muscle fatigue but also injury risk [21, 23].
4. Additive Manufacturing—A 3D printing-based manufacturing process that facilitates the creation of personalized geometries. This technology avoids rigid moulds, reduces material waste, and allows on-demand production while decreasing energy consumption from large-scale equipment and storage facilities [24, 25].
5. Advanced Polymeric Materials—Selection of high-performance polymers or thermoplastic resins, due to combined elasticity, resistance, and impact absorption properties. When integrated with generative design, these materials balance flexibility and structural support, ensuring comfort for the wearer [24, 25].

These parameters informed the design choices, ensuring that the resulting components provided adequate support without causing excessive plantar pressure, thereby optimizing overall comfort and performance [21, 26].

3.3 Unit Cells and Lattice Structures

Unit cells are the smallest repetitive units of a structure, defining the material's mechanical, thermal, and functional properties. Their configuration can be adjusted to modulate structural behaviour according to specific project requirements [15, 27].

The three-dimensional repetition of unit cells gives rise to lattice structures and is mathematically controlled by specific software, where parameters such as geometry and density can be modified. This level of control allows the optimization of material distribution and the adaptation of the structure to local mechanical requirements, improving overall performance [15, 28]. However, the classification of these structures is not universally standardized and may be based on criteria such as periodicity, topology, homogeneity, and geometry, as shown in Fig. 3.

Periodicity in lattice structures can be classified into three main categories. Periodic structures feature uniformly sized unit cells arranged in a regular pattern, ensuring predictable and repetitive mechanical properties. In contrast, pseudoperiodic structures vary the size of unit cells across the structure, allowing localized adaptation in regions that require greater resistance or flexibility. Meanwhile, random structures do not follow a fixed pattern, resulting in non-uniform and less predictable properties, favouring stress dissipation [28, 29].

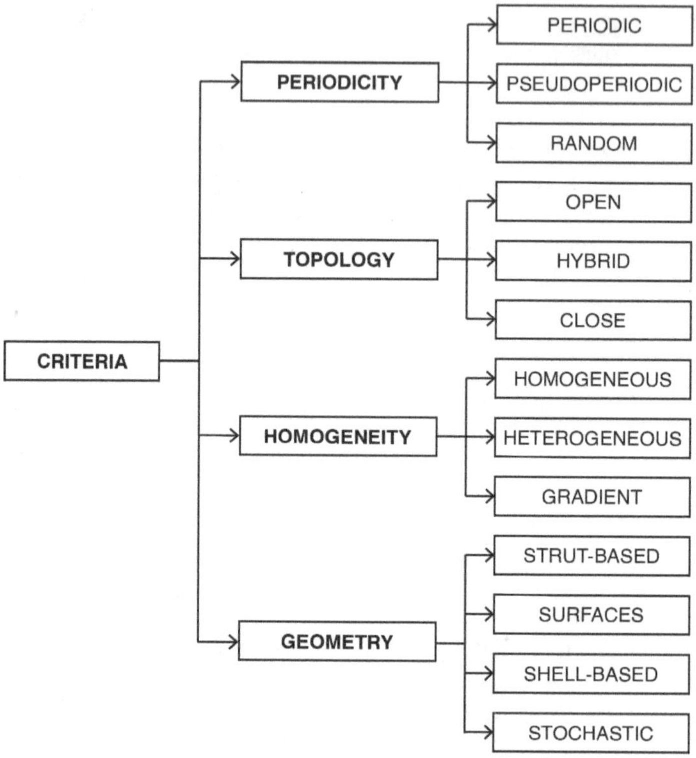

Fig. 3 Classification criteria for unit cells and lattice structures

From a **topological** perspective, the structures can be categorized as open, closed, or hybrid. Open structures lack some of their faces, making them lighter but less resistant. Closed structures have all faces sealed, maximizing structural stiffness and thermal insulation. Hybrid structures combine characteristics of both, offering a balance between lightness and strength [30].

In terms of **homogeneity**, lattice structures can be classified as homogeneous, heterogeneous, or gradient-based. Homogeneous structures maintain uniform thickness, ensuring even stress distribution. Heterogeneous structures incorporate thickness variations, which contribute to material adaptability and flexibility. Gradient structures represent a gradual transition between homogeneous and heterogeneous configurations, balancing rigidity and flexibility across different regions [28].

Another key classification method is **geometry**, which includes:

- Strut-based lattices—Composed of interconnected beam-like supports, where density and thickness can vary to control strength and stiffness. These structures exhibit high stiffness and resistance but have limited adaptability to dynamic loads [28, 31].

- Surface-based lattices—Consist of continuous surfaces that efficiently distribute stress while offering high strength-to-weight ratio. This category includes skeletal structures, where the internal volume is limited to a minimal surface, resulting in compact structures dominated by flexural behaviour, and sheet-based structures, created by thickening the minimal surface, making them better suited for absorbing tensile or compressive loads. These structures are recognized for excellent stress distribution and impact absorption capabilities [32–34].
- Prismatic lattices—Defined by cross-sections that remain constant along a single direction, producing configurations similar to honeycombs. These are effective for loads applied in a single direction, but the low geometric complexity makes them unsuitable for dynamic or multidirectional loads [28, 31].
- Stochastic lattices—Characterized by a random distribution of structural elements, providing high energy dissipation, dynamic adaptability, and resilience against failures. However, they exhibit lower mechanical predictability during the design phase, complicating structural parameter optimization [28, 35, 36].

The different classification types result in a wide range of unit cell variations, demonstrating the complexity and flexibility that generative design offers in developing optimized products. The classification criteria served as a reference for the comparative analysis conducted in this work, supporting the evaluation of structural potential in terms of comfort, adaptability, and mechanical performance.

3.4 Simulation and Structural Analysis

In this study, plantar pressure patterns are considered a key input for guiding structural decisions during the design phase. These data help define reinforcement zones and configure lattice structures according to user-specific needs. To evaluate the effectiveness of such design strategies, computational simulations and structural analysis are employed to predict how the footwear responds to forces during gait. This allows for the assessment of mechanical, thermal, and functional behaviour before production, enabling design modifications to maximize performance at the same time as reducing costs and development time [21].

Human gait is a dynamic and cyclical process characterized by the continuous weight transfer between the legs, varying the pressure exerted on different foot areas throughout each step. During this movement, changes in the plantar force distribution occur, revealing individual biomechanical patterns, including deformities and postural imbalances [37, 38]. Typically, the initial impact is registered at the heel, followed by a progressive shift of pressure toward the toes during the propulsion phase. However, variations in plantar pressure distribution can be observed depending on foot type. Individuals with *pes cavus*, high arch, exhibit increased pressure on the heel and forefoot, with reduced contact in the midfoot region, potentially leading to discomfort and instability. In contrast, those with *pes planus*, flat feet, demonstrate a more even distribution of pressure across the sole, with higher loads in the midfoot and reduced arch support, which may contribute to fatigue and overuse injuries. Neutral foot alignment presents a balanced transition of forces, optimizing gait efficiency and reducing strain on specific foot regions [39, 40].

Finite Element Analysis is a commonly used method for evaluating structural behaviour, as it facilitates analysis of stresses, deformations, and load distribution under defined conditions. Within this scope, static scenarios were considered, where the control conditions approach typical loading situations. Constraints correspond to fixed areas, such as contact zones with the ground, while loads represent external forces, including the user's weight. Although this approach doesn't capture full gait dynamics, it provides insights into mechanical performance and structural response [21].

The simulation begins with converting the digital model's geometry into a computational mesh, subdividing the structure into finite elements that enable localized calculations. The most appropriate analysis method for this application is nonlinear analysis, as it considers large displacements and contact effects between surfaces, providing a more precise prediction of behaviour under load [32, 41]. Simulation results can be interpreted through displacement diagrams, which illustrate the material's response to applied forces and stress–strain graphs that reveal variations in mechanical stress. Additionally, the safety factor is assessed to determine whether the structure can withstand real-world conditions without failure. This process helps identify spaces of excessive compression, allowing for iterative refinements that correct structural weaknesses and enhance performance until the desired criteria are met [32, 42].

Although physical validation was beyond the scope of this work, the simulated responses provided a basis for comparing structural behaviour, helping to identify areas prone to excessive compression or insufficient support.

In future stages, physical testing is necessary to validate the simulation results and ensure that the actual behaviour of the footwear aligns with numerical predictions. At this point, plantar pressure distribution can be reassessed to evaluate improvements in comfort and injury prevention [42, 43]. Integrating these insights into the development process makes it possible to redistribute pressure more efficiently, reinforce critical zones, and enhance impact absorption, contributing to better biomechanical alignment [21, 44].

3.5 Personalization

Personalization in the footwear industry allows models to be adapted to individual needs, addressing both aesthetic preferences and ergonomic or orthopaedic adjustments, particularly for pathological conditions that affect gait and comfort [45, 46].

This research analysed structural personalization from a functional perspective, focusing on how different unit cell configurations can meet specific user demands. Given the diversity of available geometries, selecting the most appropriate structure is essential to ensure a proper balance between support, flexibility and comfort. To guide this selection, factors such as activity type, intended functionality, required support, and materials were considered [47].

The selected footwear is intended for low-impact activities, primarily casual use, where comfort and lightness take priority regardless of the lattice structure applied. To achieve this, the design must balance rigidity, flexibility, and impact absorption, ensuring adequate protection while adapting to individual needs. High-pressure areas, such as the heel and forefoot, require reinforcement structures to withstand greater loads, while low-demand regions can incorporate a more porous and compliant configuration. Additionally, material selection determines manufacturing feasibility and affects properties, including elasticity, resilience, and durability [21, 37].

Among the various structural configurations, surface-based lattices are particularly effective in distributing stress evenly and promoting overall comfort, concurrently, stochastic lattices offer advantages in impact absorption and resilience [28, 36, 48]. These options were evaluated based on comfort, flexibility, and mechanical response, which informed the comparative criteria used in this analysis.

Despite that, studies indicate that single-layer lattices are more sensitive to thickness variations, while multilayer or sandwich structures offer better control over mechanical behaviour. Multilayer designs consist of overlapping layers that locally modulate material properties, increasing stiffness in support areas and flexibility in comfort zones. This structure helps prevent localized stress concentrations, making it ideal for pressure redistribution. Conversely, sandwich configurations combine two rigid outer layers and a softer or porous core, contributing to structural stability by absorbing impact and reducing internal stresses [32, 47].

Table 1 presents a comparative summary of the different structural types, highlighting higher and lower impact zones and the most suitable footwear applications for each configuration.

4 Software

As part of the methodological approach, a comparative analysis was conducted to evaluate the capabilities of selected software in supporting generative design. The analysis focused on VPLs, chosen for their intuitive data manipulation and the availability of free or licensed tools suitable for geometry creation and optimization.

The goal was to identify platforms capable of generating structures adapted to foot anatomy, balancing impact absorption, weight reduction, and structural strength. Considering this, integrating generative design with finite element analysis in a single environment offers a clear advantage by enabling a continuous workflow [29].

Thus, software with parametric modelling and structural optimization capabilities was analysed, including FreeCAD, nTop, PTC Creo, Rhinoceros 3D, and SolidWorks. To facilitate interpretation, the comparative assessment is structured by software based on lattice structures and simulation workflows, detailing how each platform aligns with the following criteria:

Table 1 Comparison of structural configurations for footwear applications: description, impact areas, and recommended usage

Structure	Description	High-impact areas	Low-impact areas	Recommended footwear type
Surface lattices	Distribute tension evenly, providing overall comfort	–	Entire foot surface	Daily use and prolonged comfort
Stochastic lattices	Better impact absorption and resilience	Heel and forefoot	Plantar arch	Walking and recreational activities
Single-layer structures	Sensitive to thickness variations, lower mechanical control	–	Entire foot surface	Lightweight and flexible
Multilayer structures	Adjust rigidity and flexibility according to the area, redistributing pressure	Heel and forefoot	Plantar arch and foot sides	Orthopaedic and sports
Sandwich structures	Rigid faces for stability, porous core for impact absorption	Heel and forefoot	Plantar arch	Performance, orthopaedic, and sports

- Adaptation to inputs—Adjusting the design according to specific data parameters.
- Generation capability—Creating geometries for unit cells and lattice structures.
- Material definition—Providing material libraries or allowing the insertion of custom material definitions that simulate real-world behaviour.
- Boundary condition configuration—Applying loads, constraints, and parameters to ensure adaptation to actual usage conditions.
- Analysis capability—Providing structural and mechanical evaluation resources, including stress–strain mapping.
- Parametric optimization—Adjusting density, thickness, and topology to achieve improved results for each footwear area.
- Compatibility with additive manufacturing—Supporting the export of models in 3D-printable formats, thereby eliminating manual post-processing steps.

Table 2 presents a comparative summary of the analysed software, categorizing their correspondence levels into three groups: high correspondence, indicating full compatibility with the specified criterion; partial correspondence, denoting limited functionality or dependency on additional tools; and low correspondence, representing minimal or no support for the feature.

The comparative analysis of the software revealed that, despite its widespread use, SolidWorks lacks native support for unit cell generation, requiring manual operations

Table 2 Comparison of leading generative design and finite element analysis software, focusing on the defined criteria

Criteria	FreeCAD	nTop	PTC Creo	Rhinoceros 3D	Solidworks
Input adaptation	Partial	High	High	Partial	High
Generation capacity	Partial	High	High	Partial	Low
Material definition	Partial	High	High	Partial	High
Boundary condition configuration	Partial	High	High	Partial	High
Analysis capability	Partial	High	High	Partial	High
Parametric optimization	Partial	High	High	Partial	Partial
Compatible with additive manufacturing	High	High	High	High	High
Obs	Code-dependent Manual operations High learning curve	Intuitive interface Optimization and automation processes	High learning curve Optimization processes	Dependency on external plug-ins Basic operations High manual refinement	Traditional approach with limited generative design capabilities High manual refinement

for parametric optimization. Similarly, Rhinoceros 3D and FreeCAD exhibit significant limitations, including a reliance on plug-ins and the requirement for advanced programming knowledge, which may complicate practical implementation.

Among the software options analysed, nTop and PTC Creo emerge as the most comprehensive solutions, encompassing modelling and structural simulation capabilities. nTop distinguishes itself through its intuitive interface and advanced automation features, positioning it as the most efficient alternative for optimizing footwear design.

5 Conclusion

The analysis demonstrated the impact of structural approaches on footwear performance, emphasizing the relevance of generative design in product optimization. The selection of the optimal structure is influenced by variables such as mechanical properties, manufacturing complexity, scalability, and cost. In this regard, generative design presents distinct advantages, including:

- Weight reduction—Enables the design of lightweight and efficient structures by minimizing material usage without compromising mechanical strength.
- Performance improvement—Facilitates the creation of designs highly tailored to specific requirements, such as resistance and flexibility.
- Accelerated development process—Automates the generation of multiple design variants, thereby streamlining production and reducing lead time.
- Expanded design possibilities—Utilizes algorithms and parametric modelling to create shapes beyond the limitations of traditional methods.
- Greater efficiency—Integrates structural simulations into the design process, reducing resource consumption and the need for physical prototyping.
- Personalization and accessibility—Allow adaptable geometries that cater to ergonomic and aesthetic user preferences [17].

Thus, generative design distinguishes itself from conventional methods by meeting the modern demands in design and industry, fostering innovation, sustainability, and efficiency.

The next steps of this research focus on developing an optimized footwear model, beginning with 3D scanning of a human foot and mapping plantar pressure. Subsequently, material testing will determine the most suitable option, considering criteria such as mechanical performance, comfort, and manufacturing feasibility. These stages aim to advance the creation of an innovative and functional solution, contributing to the development of personalized footwear.

Acknowledgements This study was funded by FAIST Project—Sustainable and Technological Smart Agile Factory, financed by funds from the "Next Generation EU" program of the Recovery and Resilience Plan (PRR) Ref. C644917018-00000031

References

1. DiNapoli T, Global shoe waste: the environmental impact of footwear. https://www.unsustainablemagazine.com/global-shoe-waste/, Last accessed 17 June 2024
2. Motawi W (2015) How shoes are made: a behind the scenes look at a real sneaker factory
3. Villena, Daniel S (2024) Toe injuries. In: Slullitel P, Rossi L, Camino-Willhuber G (eds) Orthopaedics and Trauma. Springer, Cham. https://doi.org/10.1007/978-3-031-30518-4_155
4. Luximon A, Ma X (2013) Design and manufacture of shoe lasts. In: Handbook of footwear design and manufacture. Woodhead Publishing Limited, Chap. 9, pp 177–196
5. Lillis C (2019) Foot bones: everything you need to know. https://www.medicalnewstoday.com/articles/324336, Last accessed 03 Oct 2024
6. Di Roma A (2017) Footwear Design. The paradox of 'tailored shoe' in the contemporary digital manufacturing systems. Design J. https://doi.org/10.1080/14606925.2017.1352780
7. Matthews D, Cryer-Coupet Q, Degirmencioglu N (2021) I wear, therefore i am: investigating sneakerhead culture, social identity, and brand preference among men. Fash Text 8:1. https://doi.org/10.1186/s40691-020-00228-3

8. Grand View Research: Sustainable Footwear Market Size, Share & Trends Analysis Report by Type (Athletic, Non-athletic), By End-user (Men, Women, Children), By Region, And Segment Forecasts, 2023–2030 (2023). https://www.grandviewresearch.com/industry-analysis/sustainable-footwear-market, Last accessed 08 July 2024
9. Egan PF (2023) Design for additive manufacturing: recent innovations and future directions. Designs 7(4):83. https://doi.org/10.3390/designs7040083
10. Van Rensburg ML, Nkomo SL, Mkhize NM (2020) Life cycle and End-of-Life management options in the footwear industry: a review. Waste Manag Res 38(6):599–613. https://doi.org/10.1177/0734242X20908938
11. Bartlett J, Atacama: O grande cemitério da moda global no deserto chileno. https://www.nationalgeographic.pt/meio-ambiente/cemiterio-moda-deserto-de-atacama-chile_4909, Last accessed 06 Feb 2025
12. Cheah L et al (2012) Manufacturing-focused emissions reductions in footwear production. J Clean Prod. https://doi.org/10.2139/ssrn.2034336
13. Ghimouz C, Kenné JP, Hof LA (2023) On sustainable design and manufacturing for the footwear industry—towards circular manufacturing. Mater Design 233. https://doi.org/10.1016/j.matdes.2023.112224
14. Alapide B Jr. et al (2024) Efficiency enhancements on footwear production line: a case study utilizing promodel simulation. In: Proceedings of the 7th European industrial engineering and operations management conference. Augsburg, Germany. https://doi.org/10.46254/EU07.20240179
15. Terazzi A (2022) Design with generative design method of motorcycle triple clamps made by additive manufacturing. Polytechnic of Turin
16. Kookalani S et al (2024) Trajectory of building and structural design automation from generative design towards the integration of deep generative models and optimization: a review. J Build Eng 97. https://doi.org/10.1016/j.jobe.2024.110972
17. McKnight M (2017) Generative design: what it is? how is it being used? why it's a game changer. KnE Eng 2(1):176–181. https://doi.org/10.18502/keg.v2i2.612
18. Chew ZX, Wong JY, Tang YH, Chieh Yip C, Maul T (2024) Generative design in the built environment. Autom Constr 166. https://doi.org/10.1016/j.autcon.2024.105638
19. Savadkoohi BT, De Amicis R (2009) A CAD system for evaluating footwear fit. In: Ślęzak D, Grosky WI, Pissinou N, Shih TK, Kim Th, Kang BH (eds) Multimedia, computer graphics and broadcasting. MulGraB 2009. communications in computer and information science, vol 60. Springer, Berlin. https://doi.org/10.1007/978-3-642-10512-8_1
20. Yuan M et al (2021) 3D foot scanning using multiple RealSense cameras. Multimed Tools Appl 80:22773–22793. https://doi.org/10.1007/s11042-020-09839-w
21. Tang Y et al (2021) Data-driven design of customized porous lattice sole fabricated by additive manufacturing. Procedia Manuf 53:318–326. https://doi.org/10.1016/j.promfg.2021.06.035
22. Issolio A (2018) High technology process innovation in footwear design. Polytechnic of Milan. https://hdl.handle.net/10589/142173
23. Amorim D et al (2019) Exploring mechanical meta-material structures through personalised shoe sole design. In: SCF '19: proceedings of the 3rd annual ACM symposium on computational fabrication, vol 2, pp 1–8. https://doi.org/10.1145/3328939.3329001
24. Trapp M et al (2022) Improving sustainability of footwear production through 3D printing of shoes. In: Series of publications by the scientific society for work and business organization, pp 1–15. https://doi.org/10.30844/WGAB_2022_1
25. Cui T et al (2024) Application of 3D printing technology in footwear design and manufacture—a review of developing trends. Textile Leather Rev 7:1304–1321. https://doi.org/10.31881/TLR.2024.151
26. Chen W-M et al (2022) Modulation of internal tissue stresses of the knee via control of variable-stiffness properties in a 3D-printed footwear: a combined experimental and finite element analysis. Med Eng Phys 104. https://doi.org/10.1016/j.medengphy.2022.103800
27. Belford R, Redes Cristalinas e Células Unitárias. https://chem.libretexts.org/@go/page/43268, Last accessed 02 Jan 2025

28. McGregor M et al (2021) Architectural bone parameters and the relationship to titanium lattice design for powder bed fusion additive manufacturing. Addit Manuf 47. https://doi.org/10.1016/j.addma.2021.102273
29. Pei E, Kabir I, Breški T et al (2022) A review of geometric dimensioning and tolerancing (GD&T) of additive manufacturing and powder bed fusion lattices. Progress Addit Manuf 7:1297–1305. https://doi.org/10.1007/s40964-022-00304-8
30. Nazir A, Hussain S, Muhammad Ali H, Waqar S (2024) Design and mechanical performance of nature-inspired novel hybrid triply periodic minimal surface lattice structures fabricated using material extrusion. Mater Today Commun 38. https://doi.org/10.1016/j.mtcomm.2024.108349
31. Veloso F, Gomes-Fonseca J, Morais P et al (2022) Overview of methods and software for the design of functionally graded lattice structures. Adv Eng Mater 24. https://doi.org/10.1002/adem.202200483
32. He L, Li T, Zhong D et al (2023) Blast resistance in sandwich structures based on TPMS. Buildings 13(11). https://doi.org/10.3390/buildings13112835
33. Al-Ketan O, Rowshan R, Abu Al-Rub RK (2018) Topology-mechanical property relationship of 3D printed strut, skeletal, and sheet based periodic metallic cellular materials. Addit Manuf 19:167–183. https://doi.org/10.1016/j.addma.2017.12.006
34. Yeranee K, Rao Y, Triply periodic minimal surfaces thermal hydraulic effects. Encyclopedia. https://encyclopedia.pub/entry/39188, Last accessed 02 Jan 2025
35. Huang H, Wang L, Fan Y (2023) Metallic meta-biomaterials: a critical review of fatigue behaviors. J Sci Adv Mater Dev 8(3). https://doi.org/10.1016/j.jsamd.2023.100585
36. Krešić I, Kaljun J, Rašović N (2024) Controlling the mechanical response of stochastic lattice structures utilizing a design model based on predefined topologic and geometric routines. Appl Sci 14(14):6048. https://doi.org/10.3390/app14146048
37. Schmeltzpfenning T, Brauner T (2013) Foot biomechanics and gait. In: Handbook of footwear design and manufacture. Woodhead Publishing Limited, Chap. 2, pp 27–48
38. Lou C et al (2017) A graphene-based flexible pressure sensor with applications to plantar pressure measurement and gait analysis. Materials (Basel) 10(9):1068. https://doi.org/10.3390/ma10091068
39. Department of Veterans Affairs: Center for Limb Loss and Mobility. https://www.amputation.research.va.gov/limb_loss_prevention/Functional_Limb_Loss_Prevention.asp, Last accessed 07 Mar 2025
40. Buldt AK, Landorf KB, Levinger P et al (2015) Foot posture is associated with kinematics of the foot during gait: a comparison of normal, planus and cavus feet. Gait Posture 42(1):42–48. https://doi.org/10.1016/j.gaitpost.2015.03.004
41. Zhmaylo M, Maslov L, Borovkov A et al (2023) Finite element homogenization and experimental evaluation of additively manufactured lattice metamaterials. Braz Soc Mech Sci Eng 45:299. https://doi.org/10.1007/s40430-023-04210-0
42. Patel BH, Pandey PM (2024) Comparison of energy absorption characteristics of novel lattice structures inspired by Helleborus petticoat flower and fish scale pattern for different loading orientations. Thin-Walled Struct 205. https://doi.org/10.1016/j.tws.2024.112542
43. Álvarez-Trejo A, Cuan-Urquizo E, Bhate D, Roman-Flores A (2023) Mechanical metamaterials with topologies based on curved elements: an overview of design, additive manufacturing and mechanical properties. Mater Design 233. https://doi.org/10.1016/j.matdes.2023.112190
44. Hou H, Yue Y, Liu J, Xi D, Liu S (2022) Numerical simulation and experimental study of the stress formation mechanism of FDM with different printing paths. J Renew Mater 11(1):273–289. https://doi.org/10.32604/jrm.2022.021167
45. Pallant JL, Karpen IO, Sands SJ (2022) What drives consumers to customize products? The mediating role of brand experience. J Retail Consum Serv 64. https://doi.org/10.1016/j.jretconser.2021.102773
46. Keyser AV (2015) Sneaker century: a history of athletic shoes. Lerner Publishing Group, Inc
47. Nickerson KA, Li EY, Telfer S et al (2024) Exploring the mechanical properties of 3D-printed multilayer lattice structures for use in accommodative insoles. J Mech Behav Biomed Mater 150. https://doi.org/10.1016/j.jmbbm.2023.106309

48. Li Z et al (2024) Inverse design of cellular structures with the geometry of triply periodic minimal surfaces using generative artificial intelligence algorithms. Eng Struct 321. https://doi.org/10.1016/j.engstruct.2024.118988.

Design and Development of an Interface for Personalized Footwear: First Insights

Margarida Graça, Miguel Terroso, and Nuno Martins

Abstract This research focuses on the interface design of a digital platform dedicated to footwear personalization. The interface solution under study aims to enable consumers to personalize their footwear, from foot scanning to aesthetic personalization of the product. It also seeks to provide intuitive navigation as well as an engaging and satisfying shopping experience. This article presents the results of the first research phase, which is dedicated to assessing the problem; reviewing the literature on Co-design, emerging technologies in the industry, UX/UI design, and usability heuristics; as well as defining the methodologies for developing the interface design solution. To this end, a comparative analysis (benchmarking) of existing platforms is conducted, to identify a set of requirements that will serve as the foundation for the future development of the proposed interface.

Keywords Personalization · UI/UX design · Design process

1 Introduction

This research is part of Work Package 6 (WP6) of the FAIST project, which results from a partnership between the Polytechnic Institute of Cávado and Ave (IPCA) and the Footwear Technology Centre of Portugal (CTCP). FAIST aims to drive the transformation and repositioning of the Portuguese footwear industry through the integration of new technologies, processes, and sustainable materials, while strengthening its ability to meet market demands [1].

M. Graça (✉) · M. Terroso · N. Martins
Research Institute for Design, Media and Culture, School of Design, Polytechnic Institute of Cávado and Ave, Barcelos, Portugal
e-mail: mlgraca@ipca.pt

M. Terroso
e-mail: mterroso@ipca.pt

N. Martins
e-mail: nmartins@ipca.pt

D. Raposo et al. (eds.), *Strategic Design for Sustainable Futures*, Springer Series in Design and Innovation 65, https://doi.org/10.1007/978-3-032-33282-0_4

Developed within the FAIST project, this research focuses on the design and development of an interactive digital interface that allows users to scan their feet, select a footwear model, and personalize the product aesthetically. In this context, it becomes possible to align footwear with consumers' individual needs and preferences, strengthening the connection between users and products, while enhancing the perceived value of purchased goods. Thus, the proposed solution will play a crucial role in the user experience, guiding them through all stages of the process, from foot scanning to footwear customization, enabling the selection of colours, materials, and finishes. In addition, the interface aims to ensure intuitive and efficient navigation while providing an optimized shopping experience.

This article presents the conclusions of the first research phase, including a literature review on Co-design, emerging technologies in the industry, UX/UI design, and usability heuristics, as well as a comparative analysis (benchmarking) of the interface design of leading digital footwear personalization platforms, namely, Zellerfeld, SizeRight, Volumental, Nike by You, Design Italien Shoes (DIS), and Acquarell Shoes. The article is structured into five main sections: Sect. 2 presents a description of the problem; Sect. 3 describes the state of the art in key areas of the project; Sect. 4 describes the benchmarking of digital platforms for footwear customization; and Sect. 5 presents preliminary conclusions and recommendations for future work.

2 Personalization in the Footwear Industry

2.1 Personalization and Co-Design

In the study of personalization, understanding Collaborative Design, or Co-design, has proven crucial. This participatory approach integrates consumers into the process of creating products, services, and other solutions [2, 3], allowing their needs and expectations to be translated into more precise product requirements [4]. The footwear industry is a prime example of the successful application of co-design, with brands such as Reebok, Nike, and Converse enabling consumers to personalize predefined footwear models [5, 6]. Co-Design and Mass Customization (MC) are intrinsically linked, sharing the goal of placing the consumer at the centre of the creative process [6, 7]. Co-design can even be understood as an extension of MC [6]. While MC aims to produce personalized goods and services with mass production efficiency [8, 9], co-design fosters more direct consumer collaboration in product development.

Despite its inherent advantages, personalization is a complex process that can present significant challenges, such as the need for flexible and expensive technologies, difficulty in collecting detailed consumer preferences, logistical challenges in the efficient delivery of personalized products, and customers' limited willingness to bear additional costs [4]. According to Boër and Dulio [5], personalization can be categorized into three dimensions: style/aesthetic, which allows consumers to configure the visual elements of footwear without altering its structure; fit/comfort, which involves adapting the shoe's size and shape to the user's characteristics; and functionality/performance, which adjusts footwear to specific usage conditions, such as sports or professional activities.

This project focuses on adapting footwear to the consumer's foot anatomy, while also allowing for the personalization of aesthetic aspects. Therefore, according to Boër and Dulio's [5] categorization, this study addresses both fit/comfort and style/aesthetic personalization.

In recent years, innovative technologies have emerged to overcome the limitations associated with personalization and to expand their large-scale applicability. In this context, exploring these solutions and assessing their potential to drive innovation in footwear production is essential.

2.2 *Personalization and Associated Technologies*

This research highlights the relevance of technologies such as digital interfaces, 3D printing, and 3D scanners in enhancing personalization and making the production process more efficient. In the footwear industry, digitalization has democratized access to digital interfaces and interactive design tools, enabling consumers to participate actively in product creation [10]. Product configurators, that is, digital platforms that enable the modification of footwear components and attributes, play a central role in this process, tailoring products to individual preferences and enriching the shopping experience by making it more engaging and distinctive [11, 12]. Simultaneously, 3D printing has transformed footwear production and personalization [13]. Initially adopted for prototyping, 3D printing is now an efficient and versatile alternative to traditional manufacturing methods [14, 15]. Beyond enabling a high degree of experimentation in design, such as creating textures, ornaments, and personalized details, 3D printing offers significant ergonomic advantages, allowing the production of components with a level of comfort that is difficult to achieve through conventional processes [14, 15]. Today, brands such as Nike, Adidas, and Reebok have already integrated this technology into their production processes [16]. 3D scanners complement this approach by enabling precise digitization of objects and anatomies, resulting in products that are better adapted to consumer characteristics [15, 17]. In addition to improving efficiency, this technology helps to reduce waste and promotes sustainable production practices.

With the evolution of technology, these practices are expected to become more accessible and sophisticated, establishing themselves as competitive factors in constantly evolving markets [18].

3 State of the Art

3.1 Consumers and Digital Media

Consumers' behaviours and expectations. Consumers increasingly value products that not only meet their specific needs but also allow them to express their identities [19, 20]. In this context, personalization emerges as a process that actively involves consumers in product design, requiring their time and effort to define its characteristics [21]. This direct participation strengthens the emotional connection with the object, fostering a sense of ownership and belonging. Furthermore, it provides symbolic and emotional benefits, such as pride of authorship, associated with the satisfaction of creating an artifact that positively reflects their skills [21, 22]. This affective dimension is thoroughly examined within the framework of Emotional Design, which investigates how products can evoke enduring emotional responses in users [23].

Although personalized products often present higher costs, many consumers perceive that the additional expenses are justified by the products' unique and exclusive qualities [24]. This appreciation for personalized options has given rise to a hybrid consumer profile, one that is more demanding and frequently shifts between factors, such as quality, cost, and brand identity. Moreover, the demand for personalized products transcends generational boundaries, appealing to a wide range of audiences [25].

However, co-creation and personalization have introduced new challenges. On the one hand, consumers develop higher expectations regarding the quality, exclusivity, and level of detail of personalized products than mass-produced ones [26]. On the other hand, the personalization process must be supported by intuitive tools and interfaces that prevent frustration and ensure that consumers can efficiently complete creative tasks [22, 27]. In this regard, interface usability plays a crucial role in the success of personalization, directly influencing the user experience and perceived value of the final product.

Based on this perspective, an analysis will be presented on the influence of digital technologies on personalization, exploring strategies for developing interfaces that facilitate footwear co-creation.

Digital Influence in Personalization. Digitalization has been redefining consumer and production methods, making personalization a more accessible and efficient practice [6]. In addition to strengthening the relationship between brands and consumers, this evolution drives continuous innovation in the design and marketing processes, challenging traditional industry practices [28]. By providing high levels of interactivity, digital platforms allow consumers to exercise more direct control over products, leading to shopping experiences that are more engaging and personalized. To maximize the potential of personalization, brands must simplify the digital tools used, ensuring that the collaborative process is intuitive and engaging [27, 29].

In this context, Oliveira and Cunha [6] developed the Platform-User Relational Analysis Model (PURAM), a model that identifies essential dimensions for effective digital interfaces for footwear personalization. PURAM is divided into two main areas: the first covers the dimensions and categories that comprise the digital interface, and the second analyses user perceptions and behaviours. Each dimension directly impacts the user experience, affecting how users interact and make decisions during the personalization process [6]. In this context, the five dimensions, detailed and subdivided into various categories of analysis, include: Entertainment, which captures user attention and promotes interactive and engaging experiences, considering factors such as involvement level, interactivity, and the overall perception of the experience offered; Organization, which influences usability and ease of navigation, determined by elements such as layout, functionality and usability, environmental conditions, and the overall perception of the platform's structure; Informativeness, which assesses the quality of available information, ensuring that it is accurate, relevant, and up-to-date. It covers the quality of graphic elements such as icons, texts, images, and videos, as well as the overall perception of the utility and clarity of the information; Trust is directly associated with platform security and credibility, including the perceived security, conditions of use, customer support, and overall trust in the system; and Convenience, related to the ease of purchasing and accessing the product or service, influencing consumer decisions. This dimension includes the simplicity of the purchasing, delivery, and return processes, as well as the overall perception of the convenience provided.

The analysis of PURAM aims to contribute to the optimization of the usability and effectiveness of the footwear personalization interface of this project, guiding the development of a more intuitive, efficient, and user-centred solution.

3.2 Emotional Design

Emotional Design strategically incorporates affective elements into the creation of products, interfaces, and experiences, aiming to trigger specific reactions and foster meaningful connections with users. Based on principles of psychology, neuroscience, and design, Emotional Design seeks to create products that are not only functional but also emotionally engaging [30, 31].

Donald Norman, one of the leading theorists of this approach, identifies in his book Emotional Design: Why We Love (or Hate) Things three levels of emotional processing that influence user perception. The visceral level refers to the immediate and unconscious emotional response triggered by a product's appearance, playing a crucial role in market differentiation through elements such as colours, shapes, and evocative styles that create emotional associations and reinforce brand identity. The behavioural level relates to the product's usability and functionality, ease of use, and task efficiency. Fluid and intuitive interactions generate positive emotions, whereas barriers and frustrations result in negative experiences. Finally, the reflective level involves a conscious evaluation of the value and impact of the experience, considering the product's influence on self-image, social status, and symbolic value. At this level, users may tolerate functional shortcomings if the product provides emotional or social benefits [32, 33].

The importance of Emotional Design in digital experiences has been widely recognized, as it is considered an essential component of User Experience Design (UX) [30]. The integration of emotional elements in design enhances usability perception, strengthens the connection between the user and the digital product, and makes the experience more engaging and memorable [33]. Additionally, it can contribute to user well-being by promoting feelings of calmness, confidence, and satisfaction throughout the interaction [34].

In summary, both Emotional Design and UX Design share the common goal of creating engaging and positive experiences for users. Within the scope of this research, the study of Emotional Design, complemented by UX Design practices, aims to integrate a subjective and affective dimension into the interaction between the user and the footwear personalization interface, making the experience more appealing, meaningful, and effective.

3.3 UX/UI Design

User Experience (UX) and User Interface (UI) are two distinct areas that complement each other. While UX embraces the entire user experience, UI is how that experience is achieved, referring to all the visual and interactive elements that make up the interface. As such, UX involves creating navigation flows and analysing a user's journey on a platform, ensuring that their needs, behaviours and emotions are considered when developing relevant and meaningful products [35]. Essentially, UX focuses on solving problems and optimizing interactions, thereby ensuring a satisfactory and efficient digital experience [36]. A positive UX results from the integration of seven fundamental factors: useful, usable, findable, credible, desirable, accessible and valuable. The balance between these factors determines the success of digital platforms and the creation of memorable experiences [37].

On the other hand, UI focuses on the design and organization of the digital elements that make up the interface, including menus, search fields, videos, and images [38]. In addition to anticipating possible navigational difficulties, this process ensures that the interface is accessible, intuitive, and easy to use [39]. The main objective is to provide fluid control of the interface, allowing users to achieve their goals intuitively and efficiently [39, 40]. To achieve this, UI encompasses three main areas: interactivity, which defines the behaviour and functioning of the elements that make up the interface; visual design, which covers the aesthetics of the interface, such as colours, typography, images, and logos; and information architecture, which organizes and structures the content of the interface to ensure clarity and accessibility [41]. A determining factor in UI quality is usability, which refers to the ease with which a user interacts with an interface. In practical terms, this metric assesses how well a user can accomplish a particular objective within a specific setting, both effectively and efficiently [42]. Although difficult to quantify, usability can be measured through tests with real users, allowing early detection of problems at the interface and enabling improvement. In addition, usability heuristics offer essential guidelines for developing positive experiences, ensuring that digital interfaces and platforms respond adequately to users' needs and expectations.

3.4 Usability Heuristics

In 1990, Jakob Nielsen and Rolf Molich developed a set of principles for evaluating interfaces, known as usability heuristics. Heuristic evaluation makes it possible to measure the usability of an interface, providing a quick and efficient method for identifying problems and supporting decision-making [43]. Applicable to various devices, their adoption is essential to minimize errors, guarantee effective solutions, and promote intuitive and accessible navigation [44].

In the context of this research, the selected heuristics play an essential role in creating a fluid, efficient, and frustration-free personalization experience. Their application makes it possible to develop a digital solution that offers clear feedback, adopts a familiar language, facilitates error correction, and integrates a minimalist design, while ensuring usability for different user profiles. Thus, the following heuristics play a central role in this study and are subsequently applied in the benchmarking analysis, making it possible to collect and compare data between competing platforms:

- "Visibility of System Status"—The design should keep users informed about the current state of the system through appropriate feedback, allowing them to make informed decisions [44, 45];
- "Match Between the System and the Real World"—The system should use language that is familiar to the user, creating a clear connection with their everyday experiences, making the interface more intuitive, easier to understand and use [46];
- "User Control and Freedom"—Users must have autonomy over their actions and the possibility of reversing errors without complications [44, 47];
- "Flexibility and Efficiency of Use"—The system must be flexible, adapting both to new users, who need clear guidance, and to experienced users, who are looking for quick and straightforward methods of accomplishing their tasks [48];
- "Aesthetic and Minimalistic Design"—The interface should present only the essential elements, ensuring that the information is relevant to the user. In addition, a minimalist design strengthens brand identity and credibility, positively impacting public perception [44, 49];
- "Help Users Recognize, Diagnose, and Recover from Errors"—Error messages should be clear, explaining the cause of the problem and presenting easy-to-implement solutions [44].

The presented heuristics are fundamental tools for evaluating usability, facilitating the identification and resolution of problems in digital interfaces, and contributing to the development of products that effectively meet users' needs and expectations.

4 Benchmarking: Digital Platforms for Personalized Footwear

The benchmarking procedure aims to analyse the digital platforms of relevant competitors and gather information on user experience when interacting with these interfaces. From this analysis, it becomes possible to identify good practices that can be integrated to optimize usability and respond effectively to user needs.

Currently, many platforms offer footwear personalization solutions to consumers. However, it has not been possible to identify a platform that fully corresponds to the concept to be developed. Therefore, with the aim of analysing comfort personalization, we considered the SizeRight [50] and Volumental [51] platforms, both recognized as leaders in 3D scanning technologies applied to the footwear sector.

In the field of aesthetic personalization, the Nike by You platform [52] was assessed due to Nike's global standing in the footwear sector. Design Italian Shoes (DIS) [53], an Italian brand that excels in the personalization of luxury footwear, was also considered. Additionally, Acquarell Shoes [54], a Portuguese brand recognized for its cultural and geographical proximity, was selected.

To implement a strategy that combines both aesthetic and comfort personalization, the Zellerfeld platform [55] was analysed. This platform is distinguished by its production method, which is based entirely on 3D printing, making it a relevant example for this research.

The strategic selection presented makes it possible to carry out a comprehensive comparative analysis of different approaches to footwear personalization in national and international contexts.

Tables 1 and 2 present an analysis of the functionalities provided by the platforms as well as an evaluation based on usability heuristics.

Regarding the analysed platforms, it was concluded that aesthetic personalization is often associated with the creation of a personal account. Additionally, these platforms inform consumers about the brand's projects, with only one—Zellerfeld—utilizing animations and explanatory videos for this purpose. Another common element is the use of 3D models of the products, enabling real-time modifications, 360° visualization, and zoom, which enhances personalization and interaction between the products and their consumers. Some platforms such as Nike by You and Acquarell Shoes offer design suggestions to users. However, only Nike by You provides an overview of the personalization process, informing the user of how many steps need to be completed.

Immediate feedback on user choices and fluid navigation are predominant features of all platforms. Across the board, there is the adoption of a minimalist design and

Table 1 Comparative analysis of the functionalities of shoe personalization platforms. Source: Own formulation, 2025

PLATFORMS	SizeRight	Volumental	Nike by you	DIS	Acquarell shoes	Zellerfeld
Login/Logout			X	X	X	
Personalization of style/ Aesthetics			X	X	X	X
Fit/Comfort personalization	X	X				X
Search filters			X			X
Visualization with real-time changes			X	X	X	
360° view and zoom			X	X	X	
Viewing price variations				X	X	
Automatic design suggestions			X		X	
Overview of the personalization process			X			
Explanation of the project/ Products			X	X	X	X
Animations/ Explanatory videos						X
Error messages	X	X			X	

clear backgrounds, highlighting products and images in contrasting colours, which demonstrates a concern for the perception of information and clarity in reading. On platforms that include a purchase option, there is clarity in the product descriptions and a simple and efficient purchase flow.

According to the Platform-User Relational Analysis Model (PURAM), it was observed that the four aesthetic personalization platforms respond positively to the established dimensions. Thus, it is essential to ensure a clear and intuitive perception of the information presented to users during the footwear personalization process. Any ambiguity in the representation of the available options or the visualization of the final product may compromise the personalization experience, affecting decision-making and, consequently, user satisfaction.

Table 2 Comparative analysis based on the heuristics studied. Source: Own formulation, 2025

Platforms	Aetrex	Volumental	Nike by you	DIS	Acquarell shoes	Zellerfeld
"Visibility of System Status"	X	X	X	X	X	X
"Match Between the System and the Real World"	X	X	X	X	X	X
"User Control and Freedom"	X	X	X	X	X	X
"Flexibility and Efficiency of Use"						X
"Aesthetic and Minimalist Design"	X	X	X	X	X	X
"Help Users Recognize, Diagnose, and Recover from Errors"		X	X		X	

5 Conclusion

This study presented the first research phase of the interface design of a digital platform dedicated to footwear personalization. In this sense, identifying the main challenges and opportunities of personalization, as well as understanding the role of design in organizing and presenting relevant information, proved to be fundamental at this stage of the project's development. To recognize good practices, the main references of platforms dedicated to personalization were analysed, showing the need to guarantee a clear and intuitive perception of the information presented during navigation on the interface. Any ambiguity in the representation of the content can compromise the experience, making it difficult to scan the foot, personalize the shoe, and make a purchase, directly impacting user satisfaction.

In summary, the main objectives of this first research phase were to explore the potential of personalization in the footwear industry, establish a global state of the art framework, and develop a benchmarking analysis. From the research conducted, a set of essential features emerged for the development of an efficient, user-centred interface:

- Personal profile area - improves the browsing and purchasing experience, making the process faster, more personalized, and convenient;
- Explanation of the project and the products - communicates the brand's values and the characteristics of the products, strengthening the relationship of trust with users;

- 360° visualization of the product, with zoom and real-time changes - enriches the interactive experience, allowing the user to visualize the final product with precision and increasing confidence in the personalization process;
- Price updates depending on the personalization made - ensure transparency in costs, preventing surprises at the end of the purchase;
- Automatic design suggestions - inspire users, especially those who are undecided or looking for new ideas;
- Clear information on the steps of the personalization process - guides the user, ensuring a fluid and intuitive experience;
- Initial tour for new users - facilitates adaptation to the platform from first access, providing an efficient introduction;
- Minimalist, clear, and functional design - reduces cognitive load, highlighting the essential elements and ensuring an accessible and inclusive interface.

The next stage of the project will consist of materializing these guidelines into a practical solution, starting with the development of personas representing the different user profiles and the structuring of the information architecture. Usability tests will then be conducted to validate the prototype and ensure that the interface responds effectively to user needs, maximizing efficiency and personalization experience.

Acknowledgements This work was funded by FAIST Project—Sustainable and Technological Smart Agile Factory, financed by funds from the "Next Generation EU" program of the Recovery and Resilience Plan (PRR) Ref. C644917018- 00000031

References

1. FAIST: Indústria portuguesa investe 50 milhões para ser a mais moderna do mundo, https://www.faist.pt/#/home/pt-PT/news/5256, Last accessed 03 June 2024
2. Dubois L-E, Le Masson P, Weil B, Cohendet P (2014) From organizing for innovation to innovating for organization: how co-design brings about change in organizations
3. Sanders E, Stappers P (2008) Co-creation and the new landscapes of design. In: Co-design, pp 5–18. https://doi.org/10.1080/15710880701875068
4. Zipkin P (2001) The limits of mass customization. In: MIT sloan management review. Spring, pp 81–87
5. Boër C, Dulio S (2007) Mass customization and footwear: myth, salvation or reality? Springer, London
6. Oliveira N, Cunha J (2019) Co-design and footwear: breaking boundaries with online customization interfaces. Int J Visual Design. https://doi.org/10.18848/2325-1581/CGP
7. Oliveira N, Carvalho H, Cunha J (2022) Collaborative mass customization of footwear: conceptualization of a three-stage holistic model. In: Innovations in mechanical engineering. https://doi.org/10.1007/978-3-030-79165-0_21
8. Daaboul J, Da Cunha C, Bernard A, Laroche F (2011) Design for mass customization: Product variety versus process variety. CIRP Ann Manuf Technol 60:169–174. https://doi.org/10.1016/j.cirp.2011.03.093
9. Tseng M, Jiao J (2001) Mass customization. In: Handbook of industrial engineering—technology and operations management. Institute of Industrial Engineers, pp 684–706

10. Boër CR, Dulio S, Jovane F (2004) Editorial: shoe design and manufacturing. https://doi.org/10.1080/09511920412331292637
11. Sandrin E, Trentin A, Grosso C, Forza C (2017) Enhancing the consumer-perceived benefits of a mass-customized product through its online sales configurator: an empirical examination. Ind Manag Data Syst 117:1295–1315. https://doi.org/10.1108/IMDS-05-2016-0185
12. Wang Y, Tseng M (2011) Integrating comprehensive customer requirements into product design. CIRP Ann Manuf Technol 60:175–178. https://doi.org/10.1016/j.cirp.2011.03.091
13. Shahrubudin N, Lee TC, Ramlan R (2019) An overview on 3D printing technology: technological, materials, and applications. In: Procedia manufacturing. Elsevier B.V., pp 1286–1296. https://doi.org/10.1016/j.promfg.2019.06.089
14. Galluccio A (2022) Industry 4.0 in focus: the Adidas Speedfactory
15. Dİlek E., Yildirim M, Uzun M (2021) Additive manufacturing (3D printing) in technical fashion industry applications. Tekstilna industrija 69:4–15. https://doi.org/10.5937/tekstind2102004d
16. Ukobitz D, Faullant R (2021) Leveraging 3D printing technologies: the case of mexico's footwear industry. Res Technol Manag 64:20–30. https://doi.org/10.1080/08956308.2021.1864919
17. Bondar O, Chertenko L, Spahiu T, Shehi E (2024) Shoe customization in a mass-production mode. J Eng Fiber Fabr 19. https://doi.org/10.1177/15589250241239247
18. Oliveira N, Cunha J, Carvalho H (2019) Co-design and mass customization in the portuguese footwear cluster: an exploratory study. In: Procedia CIRP. Elsevier B.V., pp 923–929. https://doi.org/10.1016/j.procir.2019.04.285
19. Gupta J (2024) Sustainable Footwear design: balancing comfort, functionality and environmental impact. Int Sci J Eng Manag 03:1–9. https://doi.org/10.55041/isjem01486
20. Oliveira N, Cunha J (2018) Footwear customization: a win-win shared experience
21. Trentin A, Perin E, Forza C (2014) Increasing the consumer-perceived benefits of a mass-customization experience through sales-configurator capabilities. Comput Ind 65:693–705. https://doi.org/10.1016/j.compind.2014.02.004
22. Piller F (2004) Mass customization: reflections on the state of the concept. Int J Flex Manuf Syst 16:313–334. https://doi.org/10.1007/s10696-005-5170-x
23. Dominici G, Espejo R (2015) Cybernetics of value cocreation. https://doi.org/10.13140/2.1.2461.2642
24. Franke N, Piller F (2004) Value creation by toolkits for user innovation and design: the case of the watch market. https://doi.org/10.1111/j.0737-6782.2004.00094.x
25. EuroShoe Consortium (2002) The Market for Customized Footwear in Europe: Market Demand and Consumer's Preferences
26. Wind J, Rangaswamy A (2001) Customerization: the next revolution in mass customization. https://doi.org/10.1002/1520-6653(200124)15:1<13::AID-DIR1001>3.0.CO;2-#
27. Malthouse E, Hofacker C (2010) Looking back and looking forward with. Interact Mark. https://doi.org/10.1016/j.intmar.2010.04.005
28. Schultz DE, Malthouse EC (2017) Interactivity, marketing communication, and emerging markets: a way forward. J Curr Issues Res Advertising. 38:17–30. https://doi.org/10.1080/10641734.2016.1233152
29. Salvador F, Holan P, Piller F (2009) Cracking the code of mass customization. MIT Sloan Manag Rev 50
30. Shakoor A, the impact of emotional design in UX
31. Yusa M, Ardhana K, Putra N, Pujaastawa I (2023) Emotional design: a review of theoretical foundations, methodologies, and applications. J Aesthetics Design Art Manag 3:1–14. https://doi.org/10.58982/jadam.v3i1.308
32. Komninos A, Norman's three levels of design
33. Norman D, Emotional design: why we love (or Hate) everyday things. Basic Books
34. Interaction Design Foundation: What is Emotional Design (ED)? https://www.interaction-design.org/literature/topics/emotional-design, Last accessed 31 Jan 2025
35. Unger R, Chandler C (2024) A project guide to UX design : for user experience designers in the field or in the making. Pearson Education

36. Norman D, Nielsen J, The Definition of User Experience (UX), https://www.nngroup.com/articles/definition-user-experience/, Last accessed 03 Sept 2024
37. Interaction Design Foundation: The 7 Factors that Influence User Experience, https://www.interaction-design.org/literature/article/the-7-factors-that-influence-user-experience, Last accessed 09 Sept 2024
38. Löwgren J, Interaction Design—brief intro, https://www.interaction-design.org/literature/book/the-encyclopedia-of-human-computer-interaction-2nd-ed/interaction-design-brief-intro, Last accessed 09 Sept 2024
39. Charfi S, Trabelsi A, Ezzedine H, Kolski C (2014) Widgets dedicated to user interface evaluation. Int J Hum Comput Interact 30:408–421. https://doi.org/10.1080/10447318.2013.873280
40. Pratama MAT, Cahyadi AT (2020) Effect of user interface and user experience on application sales. In: IOP conference series: materials science and engineering. IOP Publishing Ltd. https://doi.org/10.1088/1757-899X/879/1/012133
41. Stevens E, What is UI design? A complete introductory guide, https://www.uxdesigninstitute.com/blog/what-is-ui-design/, Last accessed 09 Sept 2024
42. Interaction Design Foundation: What is User Interface (UI) Design?, https://www.interaction-design.org/literature/topics/ui-design? Last accessed 01 Aug 2024
43. Interaction Design Foundation: What is Heuristic Evaluation (HE)?, https://www.interaction-design.org/literature/topics/heuristic-evaluation? Last accessed 16 Sept 2024
44. Nielsen J, 10 Usability Heuristics for User Interface Design, https://www.nngroup.com/articles/ten-usability-heuristics/, Last accessed 03 Sept 2024
45. Harley A, Visibility of System Status (Usability Heuristic #1), https://www.nngroup.com/articles/visibility-system-status/, Last accessed 17 Sept 2024
46. Kaley A, Match Between the System and the Real World (Usability Heuristic #2), https://www.nngroup.com/articles/match-system-real-world/, Last accessed 04 Dec 2024
47. Rosala M, User Control and Freedom (Usability Heuristic #3), https://www.nngroup.com/articles/user-control-and-freedom/, Last accessed 28 Nov 2024
48. Laubheimer P, Flexibility and Efficiency of Use (Usability Heuristic #7), https://www.nngroup.com/articles/flexibility-efficiency-heuristic/, Last accessed 17 Sept 2024
49. Fessenden T, Aesthetic and Minimalist Design (Usability Heuristic #8), https://www.nngroup.com/articles/aesthetic-minimalist-design/, Last accessed 18 Sept 2024
50. Aetrex, https://www.aetrex.com/technology/tech-sizeright-mobile-app.html, Last accessed 20 Mar 2025
51. Volumental, https://volumental.com/volumental-mobile-scanning, Last accessed 20 Mar 2025
52. Nike, https://www.nike.com/pt/nike-by-you, Last accessed 21 Mar 2025
53. Design Italian Shoes (DIS), https://www.designitalianshoes.com/en/pages/how-it-works, Last accessed 21 Mar 2025
54. Acquarell Shoes, https://acquarellshoes.com, Last accessed 21 Mar 2025
55. Zellerfeld, https://www.zellerfeld.com/, Last accessed 21 Mar 2025

Joints in Sustainable Design: Strategies for Material Assemblies and Systems

Juan L. Roquette Rodríguez-Villamil, Fernando M. Alonso Pedrero, and Diego J. Caro Serrano

Abstract In sustainable design, the integration and articulation of elements through joints and assemblies are as crucial as material selection. This chapter emphasizes the pivotal role of these connections in fostering circular, adaptive, and resource-efficient design strategies within architecture and product design. Utilizing both additive and subtractive methodologies, the research illustrates how the design of joints transcends structural necessity, catalyzing both aesthetic innovation and functional adaptability. It reinterprets the "ready-made" concept from a sustainability viewpoint, showcasing how existing components can be innovatively repurposed into low-impact solutions. Drawing on pedagogical experimentation at the Universidad de Navarra, the narrative underscores the educational value of iterative prototyping and making. Joints are reenvisioned not merely as functional requisites but as critical design junctures where modularity, material intelligence, and cultural relevance merge to reshape contemporary sustainable practices. This approach not only enhances the structural and aesthetic qualities of design projects but also contributes to a more sustainable and resource-conscious design ethos.

Keywords Sustainable design · Joints and assemblies

1 To Make: Thinking with Hands

1.1 Introduction to Making

At the core of sustainable design lies the act of making [1]—this initial tactile interaction is where theories are tested and materials meet creativity. This hands-on approach not only solidifies the foundation of practical knowledge but also fosters a

J. L. Roquette Rodríguez-Villamil (✉) · F. M. Alonso Pedrero · D. J. Caro Serrano
Escuela Técnica Superior de Arquitectura, Departamento de Proyectos, Teoría y Urbanismo, Universidad de Navarra, Pamplona, Spain
e-mail: jroquette@unav.es

D. Raposo et al. (eds.), *Strategic Design for Sustainable Futures*, Springer Series in Design and Innovation 65, https://doi.org/10.1007/978-3-032-33282-0_5

deep connection between the creator and the creations. Engaging directly with materials and construction methods allows designers to transcend traditional boundaries of architectural and design practice [2], pushing towards innovative solutions that respond more directly to environmental and societal needs.

1.2 Rethinking the Connection

In the realm of sustainable design, joints and assemblies are reimagined as vital instruments of innovation rather than mere points of mechanical necessity [3]. Historically considered peripheral or merely technical, the joint becomes a locus for demonstrating principles such as modularity, reversibility, and lifecycle thinking [4]. These concepts are critical as they enable the design of systems that are not only sustainable but also adaptable and capable of evolving over time. This transformation in design philosophy is driven by escalating environmental degradation, material scarcity, and socio-economic instability [5], which demand more than just a reevaluation of resources; they call for a fundamental transformation in how materials, structures, and systems interconnect.

At the heart of this transformation lies the joint—the often-overlooked element that determines not only how individual parts come together but also how entire systems function, adapt, and endure over time. In a circular economy framework, the joint facilitates disassembly, reuse, and adaptability [6]; it becomes the hinge upon which material flows and spatial transformations depend (Fig. 1). The designer's role is expanded to that of a systems thinker, who balances between permanence and change, tradition and technological advancement, delivering solutions that are both timeless and timely.

The design of joints becomes an act of negotiation, embodying principles of modularity, reversibility, and lifecycle thinking, which are capable of fostering structural vulnerability into generative potential. This requires a reorientation of design thinking, positioning the joint not just as a functional mechanism but as an expression of a design ethos grounded in sustainability, resilience, and creativity. It advocates

Fig. 1 *Squabble* by M. Songco, S. Vitro, S. Cao, S. Choy and C. Paredes. Students' proposal for the Joint Project at the Faculty of Architecture, University of Navarra. Image courtesy of the University of Navarra

for a pedagogical and professional shift toward viewing the joint as a primary design element, one capable of unlocking new paradigms in sustainable construction and product development.

1.3 Tradition as Innovation/Learning from Craft

The wisdom embedded in traditional joinery techniques, particularly from cultures known for their meticulous woodwork like Japan, provides invaluable insights into sustainable practices [7]. These sophisticated systems, developed over centuries through empirical knowledge and practical refinement, embody a structural and aesthetic philosophy rooted in necessity, precision, and a deep respect for material behavior. Crafted without nails, screws, or adhesives, these joints rely entirely on the careful manipulation of geometry and material to achieve connections that are both strong and reversible, intricate yet robust [8].

Far from being merely historical curiosities, such traditions reveal fundamental design principles that remain highly relevant to contemporary sustainable practice. The precision and ingenuity embedded in traditional joints demonstrate how constraint—whether of tools, materials, or processes—can serve as a generative force for innovation [9]. These systems reflect a holistic understanding of wood as a living material—its grain, expansion, contraction, and structural potential—leading to solutions that are inherently durable, efficient, and beautiful.

Reinterpreting these techniques through modern digital tools such as parametric modeling and CNC fabrication [10] allows for a fusion of historical wisdom with contemporary innovation. This results in hybrid systems where old-world craftsmanship meets new-world technology, creating solutions that are both innovative and sustainable. By preserving craft heritage and reanimating it as a living source of innovation, we not only honor traditional methods but also propel them into the future, grounded in sustainability, resourcefulness, and a respectful dialogue with materials (Fig. 2).

1.4 Extended Discussion on Material Engagement

The connection between making and materiality is further explored, discussing how direct engagement with materials can lead to a deeper understanding of their properties, potentials, and limitations. This dialogue between maker and material is crucial for fostering an intuitive sense of material behavior, which is essential for sustainable design [11]. It cultivates a material sensibility and tectonic awareness that often eludes digitally driven design approaches, allowing designers to innovate from within constraints rather than in spite of them. Bruno Munari's reflections in *"What is Bamboo"* [12] exemplify this perspective, illustrating how an intimate knowledge of materials extends beyond their physical and geometric attributes to encompass

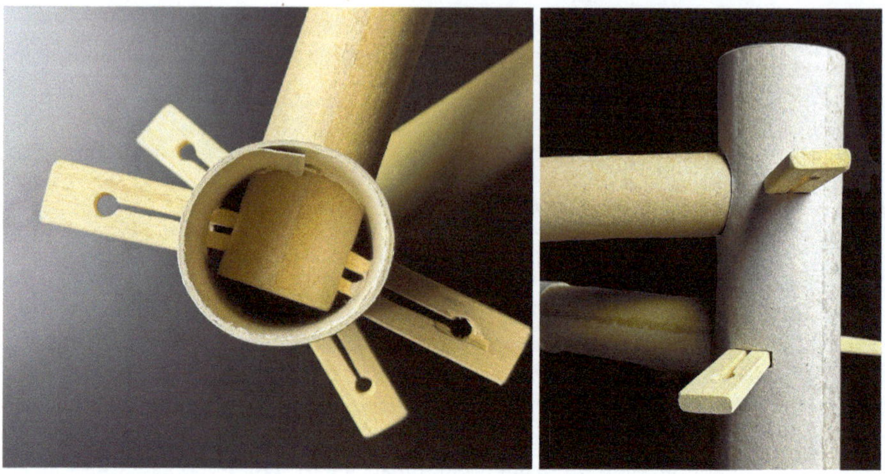

Fig. 2 *The Clip* by M. Ramirez, M. Gutierrez, F. Beverini and V. Roseros. Students' proposal for the Joint Project at the Faculty of Architecture, University of Navarra. Image courtesy of the University of Navarra

cultural values and sensorial experience [12, pp. 110–113] (Fig. 3). This expanded understanding reinforces the idea that materials are not passive elements but active participants in the design process, shaping both functional and aesthetic outcomes. Engaging directly with materials also equips students and practitioners with the ability to think critically about the physical logics of construction [13], enhancing their capability to create sustainable and responsive designs that genuinely meet contemporary needs.

By thoroughly exploring these aspects, this section not only deepens the understanding of how sustainable design is practiced but also highlights the transformative potential of rethinking conventional approaches to joints and assemblies in architecture and product design. The narrative intertwines theoretical insights with practical applications, providing a robust framework for future explorations in sustainable design innovations.

The importance of rethinking joints and material connections is further emphasized in *U-Joints: A Taxonomy of Connections* [14], a comprehensive research of how components come together in design. This work catalogues a wide array of connection typologies—from basic fasteners to advanced material junctions—revealing not only the technical ingenuity behind them but also their expressive and cultural dimensions. By dissecting and analyzing the joint as a fundamental design element, the book aligns with this chapter's aim to deepen material awareness and encourage experimentation with constructive logic. It serves as both a reference and inspiration, showing how deliberate engagement with joints can open new pathways for innovation, particularly when approached with a critical and sustainable mindset.

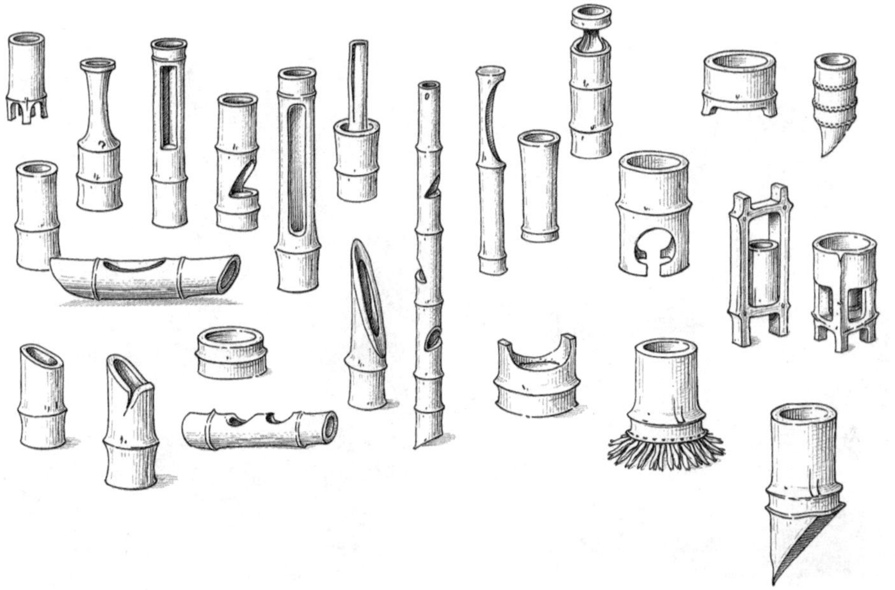

Fig. 3 Sketches presenting different bamboo working techniques, redrawn by the authors after Bruno Munari (Design as Art, 2018, p. 111). Image courtesy of the authors

2 To Measure

2.1 Introduction to Measurement

Geometry serves as the foundational framework through which designers across various disciplines and scales—ranging from product design to interior and architectural design—measure and refine the interaction between form and function, user and design. As fundamental languages of design, geometry and matter translate abstract concepts into tangible, functional systems (Fig. 4). Le Corbusier, as a pivotal figure in modern design, sought to establish a design philosophy that spanned these multiple scales and disciplines—from the design of everyday objects to large-scale architecture and urbanism. In this regard, his *Modulor* system, which bases architectural proportions on the human scale, offers a compelling example of how starting from human measurements can create designs that resonate with users on both a practical and aesthetic level [15]. The human body emerges as a universal point of reference, guiding the design of objects and spaces that are both functional and harmonious. This section delves into how these elements (geometry/dimensions/order) interact within the design process, emphasizing the essential role of measurement in developing practical and sustainable solutions.

Fig. 4 *Within the Apiary* by S. Gallardo, A. Gutérrez, P. Pavía, D. Vallejo and Y. Zhang. Students' proposal for the Joint Project at the Faculty of Architecture, University of Navarra. Image courtesy of the University of Navarra

2.2 Prototypes as Tools for Innovation

In the realm of sustainable design, prototyping is indispensable, acting as a crucial bridge between theoretical exploration and practical application. This is particularly evident in educational settings, such as at the Universidad de Navarra, where prototyping is not just about testing structural integrity but also serves as a means to engage deeply with materials through tactile and embodied knowledge [16].

Students utilize experimental models crafted from accessible, low-cost materials like cardboard, wood sticks, elastic cords, and recycled plastic. These prototypes allow them to explore balance, tension, and modularity on a manageable scale, carrying significant educational value. They simulate real-world conditions, providing students with early exposure to the challenges of design logic, material behavior, and systemic thinking.

The iterative nature of prototyping fosters a culture of inquiry and resilience, crucial in all fields of design. Discoveries made on a small scale often translate effectively to larger applications, confirming not just the structural viability but also reinforcing the underlying conceptual logic. Thus, prototyping is elevated from a mere phase to a fundamental pedagogical strategy that promotes critical thinking, ecological awareness, and hands-on experimentation. By grounding theoretical inquiry in material practice, prototyping becomes a dynamic catalyst for innovation, enabling students to visualize and critically assess the implications of their design choices while fostering a reflective approach to resource utilization, component articulation, and systemic integration [17].

2.3 Geometry and Flexibility

The role of geometry in enhancing the adaptability of design is critical across various design sectors. This adaptability is particularly crucial where the functionality and longevity of designs must respond dynamically to changing conditions. The joint, a basic yet pivotal design element, plays a central role in how modular systems like buildings, furniture, or temporary installations adjust to dynamic environments [18].

Joints evolve beyond their traditional role as mere connectors, transforming into dynamic components that facilitate load transfer, movement, and transformation within any designed space. By manipulating geometric configurations—using angles, nesting, and interlocking features—designers significantly boost the adaptability of these elements. This flexibility is essential in sustainable design, where durability and functionality depend on the ability to adapt rather than remain static.

Geometric principles allow systems to expand, contract, or reconfigure, adapting to both external pressures and internal demands [19]. This capability turns the simple joint from a static component into a strategic tool for extending the lifecycle of designs, fostering the evolution and adaptation of structures without the need for complete overhauls, thus minimizing waste and maximizing resource conservation.

The adaptability of these joints also underscores their role as enablers of resilience and sustainability within the broader context of design, ensuring that products, interiors, and buildings remain functional and relevant amidst evolving technological, environmental, and social conditions.

2.4 Geometry and Matter/Core Principles of Design

The intertwined principles of geometry and matter lie at the heart of all design endeavors, shaping the language through which design articulates its interaction with the world. These principles serve as the conduits through which design intentions are clarified, made performative, and replicable, emerging from a complex negotiation between the abstract logic of geometry and the tangible realities of materials.

Geometry provides the structural framework essential for organizing space, articulating forces, and managing loads, enabling the development of interlocking systems and modular configurations. Yet, without the tangible properties of materials—be they tactile, behavioral, or ecological—geometric designs could not transition from theory to practice.

Conversely, the variability and texture introduced by different materials enrich the design process with essential embodied knowledge (Fig. 5). Properties of materials like wood, polymers, ceramics, and cardboard directly impact the performance and adaptability of joints, pushing designers beyond theoretical constructs into the realm of practical, sustainable solutions.

This synergy is vividly illustrated in designs like geodesic domes, where the combination of spherical geometry and lightweight materials leads to efficient, scalable, and resilient structures [20]. These examples demonstrate that the progression from a single joint to an expansive spatial system is a fluid continuum, where each component, however modest, holds the potential to inspire broader innovations in design.

By emphasizing these core principles in sustainable design education, we foster an understanding of design not as an imposition of form upon material but as an ongoing, dynamic dialogue between conceptual clarity and physical reality. This perspective reimagines the joint not just as a static solution but as a vibrant, adaptable component capable of evolving in response to environmental and societal demands, highlighting its significance across all fields of design.

Fig. 5 *Vernacular* by M. Chávez, A. Evangelista, M. Ferreras, V. Muiruri and L. Vondrova. Students' proposal for the Joint Project at the Faculty of Architecture, University of Navarra. Image courtesy of the University of Navarra

3 To Gather

Transitioning from the exploration of geometry's role in enhancing design flexibility, we now explore the concept of tolerance (dimensional tolerance or 'imperfection') across various design disciplines, including product design, service design, interior design, and architectural design. Tolerance—the deliberate accommodation of dimensional variation and material inconsistency—plays an essential yet often underappreciated role in design. Far from being a flaw, tolerance is a strategic tool that introduces adaptability, resilience, and space for improvisation, enabling more responsive and innovative design solutions [21].

3.1 Tolerance, Imperfection, and Creativity

In the design of components like joints, the ability to tolerate deviation is crucial for absorbing movement, adapting to shifts in material behavior, and managing uncertainties in fabrication or site conditions. This flexibility is vital across all design fields, fostering innovative approaches. At the Universidad de Navarra, students explore how imperfection can drive creativity. Through hands-on prototyping and full-scale mockups, they confront unpredictable material behaviors—such as warping, splitting, and grain deviation—and learn to dynamically adapt their designs.

These interactions between design intentions and material realities often lead to "productive imperfections," where the discrepancies between the planned and the actual become fertile grounds for creative discovery. Here, tolerance transcends its technical role to become a conceptual strategy that celebrates dynamic engagement with materials. This approach acts as a counterbalance to the often rigid constraints imposed by digital design tools like CAD or BIM, which can stifle creativity by demanding early precision [22].

While digital tools have enabled significant advances in manufacturing, they can also result in less adaptable and more fragile design outcomes. In contrast, design processes that incorporate tangible making, prototyping, and empirical testing deepen the understanding of how components function as part of fluid, negotiable, and interconnected systems. Embracing imperfection aligns with broader philosophical and cultural attitudes toward craftsmanship, urging us to view uncertainty, error, and variability not as flaws to be eradicated but as opportunities to be harnessed.

In exploring imperfection as a method, it is essential to understand how incompleteness and imperfection play a significant role in the process of design. Richard Sennett, in his book *The Craftsman*, highlights the value of imperfections in the act of craftsmanship. Sennett contends that imperfections are not signs of failure but rather gateways to deeper creativity and innovation; imperfection, thus, invites designers to engage with their creations more intimately, fostering a deeper connection between the maker and the material. In this context, hand-drawing embodies this philosophy, where the process of sketching and conceptualization is not about achieving a flawless, final product but embracing the "tactile, relational, and incomplete nature of the creative act."

In the educational setting at Universidad de Navarra, drawing is not merely a technical skill but a method for engaging with the ideas that emerge during the early stages of a project (exploration, ideation and development). The imperfections in a drawing mirror the imperfections of the design process itself, where initial ideas often remain inchoate or fragmented. Drawing, in this sense, acts as both a medium and a method for embracing the complexity and uncertainty of creation, inviting students to engage with their projects as evolving, unfinished entities. This approach to drawing offers a compelling way to view imperfection as an opportunity for growth, experimentation, and discovery within the design process.

Ultimately, integrating precision and tolerance in design processes unveils the full creative potential of designing components, symbolizing a shift towards more humane and ecologically aware practices across all design disciplines [23].

3.2 *Incompleteness as Method/From Jazz to Joints*

The improvisational nature of jazz, where musicians start with a pre-established harmonic structure—essentially an open system—and then skillfully manipulate melodies, transforming what may appear as dissonances on the score into rich harmonic color variations, offers a powerful metaphor for design.

Just as jazz musicians transform tension into beauty through improvisation, designers in fields like product, fashion, service, or architectural design can embrace variation and even improvisation as a working method in their preliminary stages. By doing so, they can creatively navigate the unpredictability of their craft, much like how syncopated rhythms in jazz introduce dynamic alternatives to structure and timing. In both cases, the art lies not in perfect execution but in the ability to play with and transform the inherent tension into something uniquely harmonious and expressive.

Within the context of this approach to design through joints, incompleteness is not viewed as imperfection or a flaw, but as an 'open system.' This represents the system's ability to remain flexible and capable of integrating other systems or even adapting to variations of itself. This methodological shift redefines the design process as a dialogue rather than a directive, emphasizing responsiveness, intuition, and feedback from materials. When crafting components like joints, this approach might involve accommodating inherent variances in materials or adjusting designs in response to new insights. Designers thus evolve from being strict form-givers to collaborators with materials, attuned to their quirks and open to their contributions.

This philosophy is exemplified by a student project at Universidad de Navarra involving bamboo scaffolding stabilized with tensioned elastic cords. The structure was designed not to be statically perfect but kinetically adaptable, dynamically morphing and rebalancing in response to applied forces (Fig. 6). This adaptability not only demonstrated structural resilience [24] but also suggested new paradigms for thinking about design flexibility [25].

The ability of structures to dynamically adapt under varying forces also echoes natural morphogenetic processes, where form is continuously evolving, and every part of an (open) system holds potential for the whole. This biomimetic view encourages the design of components as evolving elements within living systems, not as static endpoints.

Thus, incompleteness is not a deficiency to be corrected but a valuable source of creativity and intelligence. It fosters a design culture where spontaneity and difference are catalysts for innovation, challenging the overly deterministic nature of many digital design practices that prioritize premature precision over creative freedom [26]. By drawing inspiration from jazz and organic systems, designers can cultivate a more adaptive, holistic approach to design—one that values variability, learns from diversity, and views components not just as mechanical connectors but as vibrant metaphors for connectivity, transformation, and sustainable evolution across all design fields. This active, strategic embrace of the unpredictable is key to pushing the boundaries of their practice.

Fig. 6 *Behind the Vision* by P. Aguilar, D. Alvarez, L. Ayape, N. Gallego, S. Herrera and M. Leza. Students' proposal for the Joint Project at the Faculty of Architecture, University of Navarra. Image courtesy of the University of Navarra

4 To Grow

Building on the previously discussed principles of materiality, geometry, and the interconnectivity of construction elements, the next step involves expanding design possibilities through the development of systematic approaches. Through iterative assembly and repetition, these systems reveal both physical and conceptual logics within the design process.

This section examines the practical application of these concepts in educational settings, demonstrating how they bridge the gap between theoretical knowledge and tangible design solutions. The focus shifts toward the challenges and opportunities that arise as students translate abstract ideas into physical prototypes. In this context, structural systems extend beyond their material composition to function as integral components of a broader educational framework—one that reinforces the comprehension of design methodologies while cultivating an awareness of cultural and social values.

4.1 From Joints to Systems/Educational Case Studies

The pedagogical initiative known as the *Joints Project* developed within the first year of the Design Degree curriculum at the Universidad de Navarra, functions as an experimental laboratory for testing design ideas at the intersection of material research, structural innovation, and sustainable design. This project immerses students in a hands-on, iterative design process that emphasizes the joint as a generator of larger structural systems. Within this framework, the joint is redefined not as a mere connection between components, but as an autonomous design entity—one capable of producing space, enabling assembly, and embodying principles of modularity and sustainability. As students design joints capable of replication, dissemination, and transformation, they begin to think like system designers, engaging with scalability, variability, and modular integrity. Repetition becomes a creative act—not the simple duplication of a form, but the propagation of a principle.

The *Joints Project* unfolds in three distinct but interrelated phases:

Phase 1: Exploration and Ideation—Students engage in open-ended experimentation, developing an "archive" of intuitive joint prototypes using accessible and often recycled materials. These preliminary tests prioritize direct material engagement, encouraging students to explore the potentialities and limitations of physical making without the constraints of defined formal or programmatic typologies. The result is a diverse collection of inventive joints, each reflecting a unique synthesis of structural behavior, geometric logic, and material expression.

Phase 2: Development and Implementation—This phase transitions initial explorations into full-scale working prototypes. The pedagogical emphasis shifts toward refinement, performance, and real-world viability. Students are challenged to translate abstract concepts into constructible systems, addressing stability, load transfer, tolerance, and modular repetition. The shift to a 1:1 scale is critical, as it exposes the complexities and imperfections of material behavior and fabrication processes that may not be evident at preliminary models or in digital representations.

Phase 3: Final presentation—Prototypes are curated into a public exhibition and technical presentation, where students articulate and defend their design decisions to an audience of peers, faculty, and visitors. This final stage reinforces the importance of communication and critical reflection as essential components of the design process.

The outcomes of the *Joints Project* reveal emergent typological families that highlight different strategies of connection and emphasize tectonic aspects including: mechanical joints stabilized by additional elements (e.g., bolts, clamps, or pins); friction-based systems relying on pressure and surface resistance; dry-fit joints that achieve stability through geometric interlocking; and hybrid systems that combine multiple techniques for enhanced adaptability. These prototypes operate at the convergence of ingenuity and sustainability, showing how design can flourish under constraints of material availability and environmental responsibility. Reversibility and adaptability are central to the project, as students are encouraged to develop dry joints that can be easily manipulated, adapted to different conditions, and reused.

Beyond their immediate function, these joints often act as micro-architectural systems, forming the basis for temporary pavilions, deployable furniture, or interactive installations. Thus, the project fosters an understanding of the joint as a generative module—capable of being aggregated, repeated, and transformed into spatial configurations. In doing so, students begin to internalize a systems-based logic of design, recognizing connections not as endpoints but as seeds for design growth.

4.2 Play (Fulness)

Contrary to its common association with frivolity or leisure, play in product and architectural design constitutes a serious mode of inquiry rooted in experimentation, discovery, and the suspension of rigid expectations. As Cross [27] argues, design thinking is distinct in its reliance on abductive reasoning, iterative exploration, and material engagement—an approach in which play becomes a fundamental cognitive tool. In the context of joint design, play fosters cognitive and material openness, essential to innovation. Student prototypes often emerge from playful interactions with materials and forms—exploratory gestures that evoke toys, puzzles, or constructive games. This ludic quality cultivates a sense of freedom and improvisation, allowing unanticipated relationships and patterns to emerge.

The logic of play frequently results in modular systems characterized by open-endedness, adaptability, and user-driven assembly. Several notable joint systems exemplify this approach, the most paradigmatic—and successful— example could be the *LEGO* system patented in 1958 in Denmark. LEGO, as a prototyping and educational tool, exemplifies this creative process, fostering skills such as problem-solving, innovation, and collaboration, crucial for future designers [35]. By engaging with open-ended play and iterative design, users can cultivate adaptability, encouraging a flexible approach to design thinking that transcends traditional, fixed outcomes. It's American counterpart, *K'Nex* building system created in the 1990s, comprises interlocking plastic rods, connectors, and gears. In architectural practice, Kengo Kuma's *Tsumiki* (wooden blocks in Japanese) consists of simple, triangle-shaped wooden blocks that illustrate this principle, serving both as children's building toys and scalable pavilion components, thus bridging product design and architecture (Fig. 7). Kuma's design draws from a long tradition of Japanese architecture and wood craftsmanship while also reflecting his personal explorations and childhood memories [28]. Similarly, Ronan and Erwan Bouroullec's *Algues*, a modular system of plant-like plastic units, enables a multitude of configurations, prioritizing adaptability over fixed outcomes. These systems are not designed to achieve a singular, fixed outcome; rather, they provide the rules of a game, within which users can participate, adapt, and improvise. The implications of such design logic extend beyond aesthetics: they speak to resilience, participatory construction, and the democratization of design.

A connection that begins as a game of balance or interlock can evolve into robust construction principles, refined through feedback loops of trial and error. In many cases, what starts as an "accident" in the form of an asymmetry or a misalignment, becomes a defining characteristic through repeated application. These "errors," recontextualized as features, gain value through systemic thinking and aesthetic

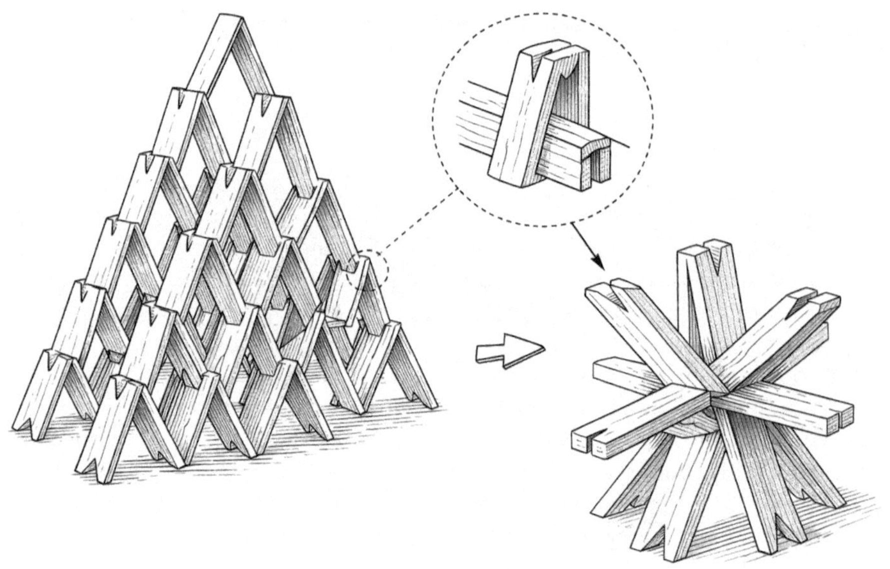

Fig. 7 *Diagrammatic representation of the Tsumiki modular system,* redrawn by the authors after Kengo Kuma & Associates. Courtesy of the authors

intention. This interplay between the iterative, hands-on work of creating individual components (tactics) and the overarching vision of a coherent, adaptable system (strategy) mirrors the relationship between design tactics and strategies as described by [29]. Tactics, such as experimenting with joint designs or testing materials, work within the framework established by the strategic direction, ensuring that the design principles are realized and refined through feedback.

The balance between strategy and tactics allows for a fluid, adaptive design process where each "error" or deviation becomes an opportunity for innovation, reinforcing both the systemic integrity and creative potential of the design. Play serves as an antidote to rigid or prescriptive design processes, introducing ambiguity, contingency, and non-linearity into the creative workflow. Within design and architectural education, fostering a culture of play not only promotes inventiveness but also encourages a tolerance for failure and uncertainty—qualities essential to adaptive and sustainable design practices. Ultimately, the culture of repetition —or the rituals— grounded in play does not inhibit creativity but it amplifies it, generating a multiplicity of outcomes from a single generative idea.

4.3 The Joint as a Social and Cultural Artifact

While joints are often understood in technical or structural terms, their significance extends far beyond the domain of engineering. Joints function as cultural objects—sites where symbolic, social, and experiential dimensions of design converge. As both physical elements and conceptual devices, they encapsulate the ways in which communities construct, inhabit, and remember their environments. They function as silent narrators of technique, tradition, and adaptation, linking the act of building to broader questions of identity, memory, and agency.

Traditional joinery techniques, such as those in Japanese carpentry or Andean adobe construction, embody generational knowledge, regional identity, and ethical values. These methods are not merely technical solutions but cultural expressions transmitted through apprenticeship and oral tradition. As Gottfried Semper proposed in *The Four Elements of Architecture* (1851) that architecture evolved from primitive needs and craftsmanship, particularly weaving. Traditional joinery, with its interwoven and interlocked components, can be understood as a translation of textile logic into timber structures, reflecting a historical evolution that continues to shape contemporary design education (Fig. 8).

Educational initiatives that frame the joint as a cultural artifact deepen students' understanding of design as a social practice. In this line, we can observe formidable contemporary extensions of this continuous acquired knowledge in Shigeru Ban's works with paper tube structures in disaster relief contexts highlighting how simplicity, modularity, and rapid assembly serve as vital tools of resilience. In these scenarios, joints must be deployable, materially efficient, and user-adaptable—factors that determine the success of humanitarian responses. Their design can make the difference between shelter and exposure, between stability and precarity. As such,

Fig. 8 *Textile Joint* by L. de la Barrera, D. Latiseva and A. Aróstegui. Students' proposal for the Joint Project at the Faculty of Architecture, University of Navarra. Image courtesy of the University of Navarra

the joint becomes an instrument of humanitarian response—facilitating collective action and empowering local communities [30].

Pivoting between the semiotic and the phenomenological, this exercise also explores the symbolic connotations of these joints through sensory experience. Students are encouraged to document material characteristics such as texture, temperature, and sound, producing catalogs that go beyond technical drawings to capture the poetic dimensions of construction. Influenced by structuralist theories, such as those of Barthes [33] and Eco [31], materials function as symbols carrying ideological meanings, which can be expanded and reinterpreted through design. Through the manipulation and re-interpretation of materials through these joint systems, their original meanings can be expanded, translated, or complemented into richer versions (Fig. 9).

By reimagining the joint as a narrative and affective artifact, students begin to see design not simply as problem-solving but as storytelling—as the creation of connections that are both physical and symbolic. This pedagogical approach expands the scope of sustainable design, moving beyond performance metrics to include cultural sustainability: the preservation and evolution of practices that root design within specific communities, ecologies, and temporalities. The joint is repositioned as a social interface where cultural heritage and technological innovation intersect. It reveals how the seemingly mundane act of connection can become a profound act

Fig. 9 *Imperfection* by A. Aleman, S. Cosío, J. Herráez, D. Osta and J. Suárez. Students' proposal for the Joint Project at the Faculty of Architecture, University of Navarra. Image courtesy of the University of Navarra

of communication—linking past and future, maker and user, material and meaning [32].

5 To (Re)make: Conclusion Toward a Sustainable Culture of Connection

The exploration of joints throughout this study affirms their central role in the construction of sustainable design—both literally, as physical connectors, and metaphorically, as conceptual frameworks for thinking about systems, relationships, and transformation. In contemporary practice, sustainability must be understood not only in terms of ecological impact or energy performance, but also in terms of **how materials are brought together, how they behave over time, and how they can be disassembled, reused, or reimagined**. Joints are at the heart of this process.

A well-designed joint encapsulates key values of sustainable architecture: **modularity**, which allows for scalability and repetition; **tolerance**, which accommodates variability and imperfection; **impermanence**, which enables disassembly and reuse; and **systemic thinking**, which fosters holistic design strategies. As pedagogical tools, joints offer a microcosm through which students can engage with a broad spectrum of architectural concerns—from structural performance to material ethics, from formal expression to cultural significance.

Moreover, the joint invites designers to embrace a more nuanced, inclusive vision of sustainability—one that values **imperfection and adaptability** over rigid precision; **collaboration and participation** over authorship; and **complexity and transformation** over static form. It challenges the notion of design as the pursuit of idealized, flawless outcomes, advocating instead for a dynamic process rooted in **experimentation, responsiveness, and embodied making**.

In this light, the joint becomes both a **technical device and a creative opportunity**—a place where innovation emerges through constraint, and where the act of connection becomes a site of possibility (Fig. 10). It is through joints that materials meet, but also through joints that ideas converge, disciplines intersect, and communities are built.

Looking forward, a sustainable culture of connection will require architects, educators, and makers to rethink how we build—not simply what we build. It calls for a renewed commitment to material intelligence, cultural empathy, and systemic imagination. In the face of global environmental and social challenges, the modest joint offers a powerful lesson: that resilience lies not in monumentality, but in the care, creativity, and curiosity with which we bring things together.

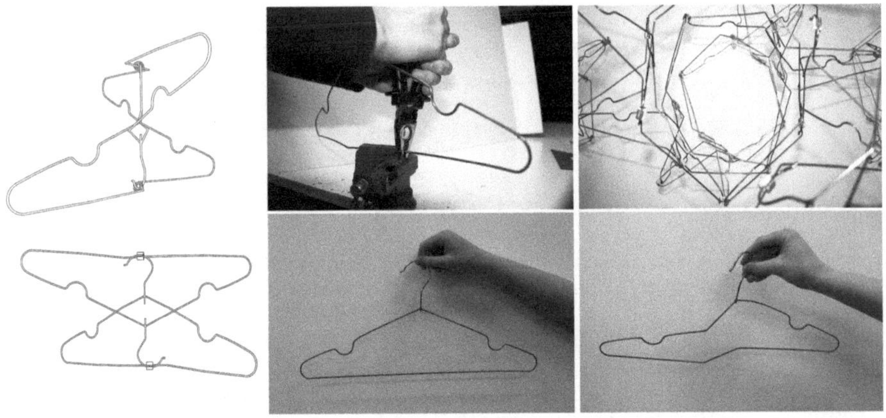

Fig. 10 *Steel* by L. Ara, M. Fernández, N. Fernández, K. Lorenzana and S. Torrens. Students' proposal for the Joint Project at the Faculty of Architecture, University of Navarra. Image courtesy of the University of Navarra

References

1. Dieleman H (2008) Cómo se puede diseñar educación para la sustentabilidad? Revista Mexicana de Investigación Educativa 13(39):1125–1148. Recuperado de https://www.scielo.org.mx/scielo.php?pid=S0188-49992008000300004&script=sci_arttext
2. Hasling KM (2018) Perspectivas del papel de los materiales en la enseñanza de diseño de producto sostenible. Temes de Disseny (34):102–113. Recuperado de https://www.raco.cat/index.php/Temes/article/download/343490/434644/
3. López M (2025) Edificios como árboles, ciudades como bosques. Editorial Sostenible, León, España
4. Cátedra de Responsabilidad Social Empresarial y Sostenibilidad (2022) Ciclo de vida del producto y productos sostenibles. Recuperado de https://catedrasostenibilidadaege.org.do/Portals/0/Documentos/CRSES/Ciclo%20de%20vida%20del%20producto%20y%20productos%20sostenibles.pdf
5. Comisión Económica para América Latina y el Caribe (CEPAL) (2020) Los motores de la degradación ambiental. Recuperado de https://repositorio.cepal.org/bitstream/handle/11362/45766/1/S2000443_es.pdf
6. Asociación Española de Normalización (UNE) (2023) Norma UNE-EN 17410:2023. Sostenibilidad en la edificación. Diseño para el desmontaje y la adaptabilidad. Principios, requisitos y directrices. Recuperado de https://www.une.org/normalizacion_documentos/Apoyo_normas_eco_circul-2024.pdf
7. Nishi K, y Hozumi K (1983) Nihon kenchiku no katachi: seikatsu no naka no tochi to kenchiku. Shōkokusha, Tokio
8. Sato H, Nakahara Y (2000) The complete Japanese joinery. Hartley & Marks, Berkeley, CA
9. Schwarz C (2008) Workbenches: from design and theory to construction and use. Popular Woodworking Books, Cincinnati, OH
10. García-Alvarado R, Turkienicz B (2021) De todos para todos: experiencias colaborativas en diseño y fabricación digital en Latinoamérica. In: Proceedings of the international conference of the association of collegiate schools of architecture (ACSA). Recuperado de. https://www.acsa-arch.org/proceedings/International%20Proceedings/ACSA.Intl.2021/ACSA.Intl.2021.184.pdf

11. Pallasmaa J (2012) La geometría de las sensaciones. Recuperado de https://tecnne.com/biblioteca/pallasmaa-la-geometria-de-las-sensaciones/
12. Munari B (2008) Design as art. Penguin Books, London
13. Papert S (1980) Mindstorms: children, computers, and powerful ideas. Basic Books, New York
14. Caputo A, Koivu A (eds) (2023) U-Joints: a taxonomy of connections. Koenig Books
15. Le Corbusier (2000) The Modulor (J. O. Urmson, Trans.). The MIT Press (Original work published 1948)
16. Soomro SA, Casakin H, Georgiev GV (2021) Sustainable design and prototyping using digital fabrication tools for education. Sustainability 13(3):1196. https://doi.org/10.3390/su13031196
17. Georgiev GV, Sánchez Milara I, Soomro SA, Casakin H, Nanjappan V (2023) Sustainable prototyping challenges in digital fabrication design education. In: 25th international conference on engineering and product design education. Elisava University School of Design and Engineering, Barcelona, Spain
18. Koronaki A, Shepherd P, Evernden M (2018) Geometry optimization of space frame structures for joint modularity. In: Mueller C, Adriaenssens S (eds) Proceedings of the IASS symposium 2018: creativity in structural design, July 16–20, MIT, Boston, USA. International Association for Shell and Spatial Structures (IASS)
19. Yang C, Xu B, Xia J, Chang H, Chen X, Ma R (2023) Mechanical behaviors of inter-module connections and assembled joints in modular steel buildings: a comprehensive review. Buildings 13(7):1727. https://doi.org/10.3390/buildings13071727
20. Maleque MA, Salit MS (2013) Materials properties and design. In: Materials selection and design. Springer briefs in materials. Springer, Singapore. https://doi.org/10.1007/978-981-4560-38-2_4
21. Ebrahim NA, Kamel SM, Hamza TS (2020) The relationship between tolerance of ambiguity and creativity in architectural design studio. Creativity Stud 13(1):179–198. https://doi.org/10.3846/cs.2020.9628
22. Attieh OSE (2019) The aesthetics of imperfection; Glitch art in three-dimension forms as a creative design tool. Int Design J 9(3). Accessed from
23. Anderson NM (2007) A case study of materiality and the necessity of craft in the age of digital fabrication. Accessed from https://www.academia.edu
24. Pawar S, Magembe B (2015) Bamboo: the emerging renewable material for sustainable construction. Constr Build Mater 81:66–73
25. Magembe B (2023) Bamboo in construction technology. In: Bamboo and sustainable construction
26. Schmitt C, Garrido FA (2022) Digital imperfection—from digital imperfection to circular perfection. In: Powerskin conference proceedings. This paper discusses the integration of digital tools
27. Cross N (2006) Designerly ways of knowing. Springer
28. Kuma K (2011) Tsumiki: Kengo Kuma's wooden blocks. Architectural J 45(7):112–118
29. Lawson B (2006) How designers think: the design process demystified, 4th edn. Architectural Press
30. Ban S, Igarashi Y (2014) Shigeru Ban: paper in architecture. Rizzoli
31. Eco U (1976) A theory of semiotics. Indiana University Press
32. Buchanan R (1992) Wicked problems in design thinking. Des Issues 8(2):5–21
33. Barthes R (1972) Mythologies (A. Lavers, Trans.). Hill and Wang (Original work published 1957)
34. Munari B (2008) Design as art. Penguin Books
35. Schiffer DR, Smith RK (eds) (2021) Cultural studies of LEGO: more than just bricks. Routledge
36. Semper G (1989) The four elements of architecture and other writings (H. F. Mallgrave & W. Herrmann, Trans.). Cambridge University Press (Original work published 1851)
37. Sennett R (2008) The craftsman. Yale University Press
38. Le Corbusier F (ed) (2000) The modulor and modulor 2. Birkhäuser
39. Rozhkovskaya N (2020) Mathematical commentary on Le Corbusier's modulor. Nexus Netw J 22:411–428. https://doi.org/10.1007/s00004-019-00469-w

40. For Designerly Practices in Geometry Cfr. Yang C, Xu B, Xia J, Chang H, Chen X, Ma R (2023) Mechanical behaviors of inter-module connections and assembled joints in modular steel buildings: a comprehensive review. Buildings 13(7):1727. https://doi.org/10.3390/buildings13071727
41. Hatch MJ (1999) Exploring the empty spaces of organizing: how improvisational jazz helps redescribe organizational structure. Organ Stud 20(1):75–100. https://doi.org/10.1177/0170840699201004
42. Mogu (2024) The value of imperfection in design: a tribute to uniqueness and authenticity. Accessed from https://mogu.bio/the-value-of-imperfection-in-design-a-tribute-to-uniqueness-and-authenticity
43. Moorman C, Miner AS (1998) Organizational improvisation and organizational memory. Acad Manag Rev 23(4):698–723. Also Cfr. Hatch MJ (1999) Exploring the empty spaces of organizing: How improvisational jazz helps redescribe organizational structure. Organ Stud 20(1):75–100. https://doi.org/10.1177/0170840699201004
44. Sharma B, Gatóo A, Bock M, Ramage M (2015) Engineered bamboo for structural applications. Constr Build Mater 81:66–73

Problem-Based Learning for Enhancing Sustainability and Circularity Concepts in Fashion Design

Teresa Raquel Barata, Marta Bicho, Rui Miguel, Ana Santiago, Sandra Ferreira, and Madalena Pereira

Abstract In a world increasingly aware of the urgent need to live in harmony with the environment, education plays a crucial role in transitioning to a more sustainable and circular model of living. The Sustainable Development Goals (SDGs), outlined in the 2030 Agenda for Sustainable Development, adopted by United Nations Member States in 2015, highlight the transformative power of education to improve the planet. They aim to equip students with the skills, critical thinking, and tools necessary to make informed choices and create sustainable solutions. This includes using more sustainable materials, ethical and preferably local production chains with lower environmental impact—particularly in the management of water, energy, and chemical products. A circular approach to fashion, with longer product life cycles and the integration of innovation and technology, is essential to preserving resources for future generations. Academia must adopt teaching practices that shift the focus from "knowing-what" to "knowing-how," aligning with sustainability and circularity strategies.

Keywords Textile and fashion · Circular economy · Problem-based learning · Sustainable fashion design

---

T. R. Barata (✉) · M. Bicho · R. Miguel · A. Santiago · S. Ferreira · M. Pereira
Fiber Materials and Environmental Technologies (FibEnTech-UBI), Universidade da Beira Interior, Covilha, Portugal
e-mail: traquel@ubi.pt

R. Miguel
CIAUD - University of Lisbon, Lisbon, Portugal

M. Pereira
UNIDCOM - IADE/UE, Lisbon, Portugal

D. Raposo et al. (eds.), *Strategic Design for Sustainable Futures*, Springer Series in Design and Innovation 65, https://doi.org/10.1007/978-3-032-33282-0_6

1 Introduction

The textile and fashion sectors are one of the most problematic industries at an environmental and social level, due to their significant contribution to climate change and intensive consumption of natural resources. The fashion industry is responsible for around 4–10% of global carbon dioxide emissions, as well as high water and energy consumption [1]. The United Nations Framework Convention on Climate Change (UNFCCC) estimates that the fashion industry contributes approximately 10% to global greenhouse gas emissions because of its lengthy supply chains and energy-intensive production processes [2]. While several gases are technically involved in greenhouse effects, carbon dioxide is by far the dominant gas associated with the fashion industry, followed by methane and nitrous oxide, primarily due to the production of plant-based materials such as cotton, which rely heavily on fertilizers, and animal-based materials such as leather and wool, which are linked to methane emissions from livestock and land use impacts. Furthermore, there is the problem of working conditions throughout the production chain, as well as the high generation of waste, both pre- and post-consumption.

Despite growing awareness of its negative impacts, the fashion industry continues to expand, driven largely by fast fashion—a model based on low-cost production, rapid turnover of consumption and the short lifespan of garments. With the rise of fast fashion, people buy cheap clothing, use them less, and discard them at an ever-increasing rate. The growth of fast fashion has contributed to a significant increase in the consumption of inexpensive clothing, which are typically worn infrequently and disposed of at an accelerating pace. Globally, over the past 15 years, the average number of times a garment is worn before being discarded has fallen by 36% [1], reflecting a significant decline in clothing use. As a result, large amounts of textile waste end up being incinerated, landfilled or exported to developing countries [3].

The fashion industry needs to reduce its carbon footprint by at least half to meet sustainability goals, but its environmental impact continues to grow. According to Sakorin and Boken [4], although the industry has focused on eco-efficiency through the circular economy, it has failed to address the core sustainability issue of over-production and over-consumption. Leal Filho et al. [5] states that the key factor in the textile decarbonization process often lies in the adoption of sustainable, low-carbon materials, coupled with advancements in manufacturing technologies. Given these concerns, it becomes imperative to promote a profound transformation in the industry's business model, based on the integration of sustainable practices throughout the supply chain and consumer re-education, promoting more conscious and responsible purchasing behaviors. Sustainability and circularity appear, in this context, as central themes in academic and professional debate, reflecting the growing awareness of the environmental and social impacts of the sector. The circular economy presents an alternative to the traditional linear model, allowing fashion designers to create products designed for durability, prolonged use, biodegradability and reduced resource consumption [6].

In fashion design, sustainable practices involve designing products that respect the environment and fair-trade principles, from choosing ethically and ecologically sourced materials to minimizing waste and emissions during production [7]. The linear system wastes economic opportunities, overloads resource, pollutes the environment and causes significant negative environmental and social impacts on various scales [1]. Circularity proposes an alternative model to the traditional linear system of "produce-use-dispose", defending the extension of the life cycle of products through reuse, recycling, reconditioning and redesign of production and consumption systems. This change requires rethinking design from its genesis, incorporating eco-design principles from the early stages of product design such as design for recyclability, modular design, multifunctionality, biodegradability and design for disassembly [3, 8]. In this scenario, training in sustainable and circular practices in the fashion sector becomes an inevitable priority. Designers today face new challenges that require a critical and systemic approach at all stages of the design process—from conception to production—contributing to the Sustainable Development Goals (SDGs) and complying with emerging regulations such as the European Green Deal. A key example is the approval of the new Eco-design Regulation by the European Council on 27 May 2024 [9], which establishes requirements regarding the durability, repairability, upgradability and reusability of products. The regulation also addresses substances that hinder circularity and sets criteria for energy and resource efficiency, recycled content, remanufacturing and recycling, including the creation of a digital product passport to promote transparency. Regulation (EU) 2024/1781 [10], known as the Ecodesign for Sustainable Products Regulation (ESPR), presents new challenges for both educators and students in the context of sustainability. Its primary aim is to facilitate the maintenance and refurbishment of products, while also enhancing the recovery of strategic and critical raw materials. Designed to reduce waste, increase the use of recycled content in products, and ensure that performance and safety standards are maintained, the regulation ultimately supports remanufacturing, high-quality recycling, and the reduction of carbon and environmental footprints.

In terms of teaching, it is necessary to develop skills in future designers that will enable them to understand and apply the principles of sustainability and circularity in the various stages of the creative development process. It has been noted that the theoretical transmission of these concepts is insufficient and that it is necessary to adopt active methodologies such as problem-based learning (PBL). The PBL methodology plays a fundamental role in the development of critical thinking, the resolution of real problems and the acquisition and deepening of sustainable practices and circularity by students [11, 12]. This methodology, when compared to traditional methodologies where the student plays a passive role, makes it possible to stimulate research skills, critical analysis and the resolution of complex problems in a creative way [13].

The application of PBL methodologies in fashion design enables the development of practical skills through the simulation of real situations with practical application. In this research, fashion design students were challenged to design a circular skirt and apply one or more sustainability and circularity practices during the development of the project. These sustainable practices included options such as upcycling, selecting sustainable raw materials and accessories, using local suppliers, optimizing patterns and cuts, as well as adopting more efficient production techniques. This approach made it possible to develop creative aspects associated with design, through the exploration of new materials, and promoted an awareness of the environmental impact of the choices made during the prototype design stages. PBL promotes experiential and integrative learning, preparing future professionals to address ethical and environmental challenges effectively. Furthermore, it encourages interdisciplinarity, by requiring students to articulate knowledge from areas such as materials science, environmental management, sociology of consumption and technological innovation reflecting the complexity of the contemporary fashion sector. Gam and Banning [14] demonstrated that PBL is an effective pedagogical approach for introducing students to sustainable design practices. By engaging with practical, real-world challenges, students were encouraged to think critically and enhance their problem-solving skills, supported by their prior design experience and strong engagement with the project. Although not directly applying PBL, Ma [15] explored the use of Challenge-Based Learning (CBL) in teaching sustainable fashion design. The CBL methodology shares core principles with PBL, such as student-centered learning and real-world problem-solving. Conducted in a graduate fashion design course, the study adopted a structured curriculum based on CBL and used a qualitative case study approach, including observational evidence, to describe students' learning experiences and outcomes. It assessed the effects on students' sustainability skills through grounded theory interview analysis and demonstrated that engaging students in real-world sustainability challenges fosters critical thinking and innovation, making CBL an effective pedagogical approach for integrating sustainability into fashion design education and improving both process and outcomes [15].

In a market increasingly oriented towards social and ecological responsibility, designers who master innovative and sustainable materials, and who understand the principles of circularity and regeneration, will be in a better position to lead innovation and transformation processes in organizations [16].

A recent report by Textile Exchange [17] shows that world fiber production reached 116 million tonnes in 2022, compared to 112 million tonnes in 2021. In the last 20 years, fiber production has almost doubled, from 58 million tonnes in 2000, and is expected to reach 147 million tonnes in 2030. This rapid growth is incompatible with the commitments made under the SDGs and the European Green Deal. In addition, the use of synthetic fibers, particularly polyester, has increased significantly, overtaking cotton as the most used fiber in the fashion industry. This is compounded by the fact that polyester is a synthetic fiber derived from fossil fuels and therefore has a higher carbon footprint than cotton. The environmental impact of the textile industry can be significantly reduced by switching to the use of more sustainable natural fibers such as organic cotton, organic wool and hemp. The use of more sustainable materials allows the textile and fashion industry to reduce emissions and the use of natural resources such as water and energy, while at the same time implementing a more sustainable supply chain [5]. Reducing the waste produced during production is one of the strategies that have been indicated for reducing emissions [5]. This waste reduction includes the reuse and recycling of pre- and post-consumer waste and the adoption of more efficient production processes. The use of new, more sustainable materials and the application of innovative design practices, based on the concepts of sustainability and circularity, will make it possible to make the fashion industry more sustainable and in line with the European Union's eco-design regulations. The incorporation of recycled materials in product development, the use of strategies such as the remanufacturing of garments and the reuse of textile accessories and materials to create new functional models is essential to reduce waste and the consumption of natural resources. The application of these practices allows for the maximization of the use of materials, reduces environmental impact, and can contribute to the development of a circular economy in fashion. The use of biomaterials and organic fibers such as organic wool, organic cotton and linen are more sustainable solutions as the production methods used involve minimal use of chemicals and reduced water consumption compared to the production of conventional fibers, which helps to reduce the ecological footprint. The fashion industry, by adopting these materials in the design process, can also create products that align with sustainability goals, reduce carbon emissions and reduce resource use, while applying concepts of circularity. In this context, it is important that fashion designers acquire knowledge and technical skills related to the use of new fibers and materials, more sustainable, to develop products with low environmental impact throughout the product's life cycle.

As stated by Bicho et al. [18], a key pillar of the circular economy strategy is waste reduction, along with the promotion of reuse, recovery, and recycling of materials. The collaboration between academic institutions and companies is equally fundamental to accelerating the transition to more sustainable materials and practices, such as the use of recyclable, biodegradable, reusable fibers or those with a lower environmental impact. These partnerships will promote experimentation and drive the integration of alternative materials and foster the development of new sustainable innovations [19]. The emergence of innovative materials, including biodegradable fabrics and recycled fibers, offers an opportunity to transition to a more sustainable textile and fashion industry.

Active teaching approaches, such as Problem-Based Learning (PBL), have proven effective for building knowledge and fostering student autonomy, especially when applied to complex and interdisciplinary topics such as sustainability in fashion design. In this context, this study aims to answer the following research question: How does the application of PBL contribute to the development of sustainability and circularity skills in undergraduate Fashion Design students? To answer this question, the following specific objectives were defined: (1) Assess the prior knowledge of students regarding the concepts of sustainability and circularity applied to fashion design; (2) Implement the PBL methodology in the development of a practical project (circular skirt) integrating sustainable and circular practices; (3) Verify at which stages of the design process students apply the concepts of sustainability and circularity; (4) Analyze the effectiveness of the PBL approach in fostering critical thinking, problem-solving skills, and the practical integration of the concepts covered; (5) Identify the main obstacles to the implementation of sustainable practices in the context of fashion design.

2 Methodology

This research aims to identify and validate teaching practices that motivate design students to develop fashion products, during laboratory sessions, using sustainable and circularity practices. It focuses on the application of hands-on learning methodologies, such as "learning by doing" and "learning to learn," within the context of textile and fashion education. The study also seeks to align with current regulations, industry demands, and student needs, and was conducted using a sample of 60 first-year Fashion Design undergraduate students.

Problem-Based Learning (PBL) methodology was used in this study to motivate students to actively engage in creating sustainable fashion products during laboratory sessions. Compared to traditional teaching methods, this methodology, by involving students in real-world practical scenarios, significantly improves learning outcomes. PBL requires students to develop practical solutions and present them in a way that develops critical thinking and problem-solving skills [20, 21]. This approach creates a learning environment that allows students to apply what they learn in the classroom to real-world contexts, leading to a deeper understanding of the topics covered. By making the connection between theory and practical application, the learning experience is more engaging and motivating for students.

Regarding the application of sustainability and circularity concepts in fashion design, the PBL methodology encourages students to critically evaluate the entire life cycle of the fashion product, assessing the importance of using durable, reusable and recyclable materials, as well as the importance of design for durability and upgradability [22]. Proper life cycle assessment and the implementation of resource cycles that aim to close the loop between post-use and production, facilitating recycling and reuse, are essential to promote sustainable practices and reduce environmental impact [22]. Through this methodology, students develop a deeper understanding of the interconnection between design, production and environmental impact.

In the first phase, students were challenged to create a sketch or drawing of a circular skirt. In the next phase, students were challenged to use sustainability and circularity practices in their projects and had to select one or more sustainable practices to apply during the development of the prototypes. These principles could include upcycling, selecting sustainable raw materials, sourcing from local suppliers and products, optimizing pattern design and fabric cutting, and adopting more efficient manufacturing techniques. At the end of the study, students presented their final projects, along with a report detailing the materials chosen, production methods and sustainable practices adopted.

In addition, a questionnaire was applied to assess students' knowledge about sustainable materials, sustainable and circular design principles and production processes, and to assess the stage at which they applied these practices in their projects. To conduct the diagnostic questionnaire, it was distributed electronically to 60 participants, yielding 55 responses. The objective was to address the following research questions:

R1—Are students aware of the difference between sustainability and circularity?

R2—Are students familiar with the practices associated with sustainability and circularity?

R3—Do students apply the concepts of sustainability and circularity in the development of new projects?

The main concepts of sustainability and circularity were introduced prior to the development of the project of a circular skirt. Table 1 summarizes the key concepts presented, which were also addressed in the diagnostic questionnaire. During the initial brainstorming session, their relevance to fashion design was discussed and emphasized. In addition to promoting the integration of sustainable and circular practices, the potential of design as a driver of behavioral change was also explored. Critical reflection on consumption habits and the issue of fast fashion was encouraged, with the aim of developing an awareness of the social role of the fashion designer. This overview aimed to guide students in making design decisions during the development of their circular skirt projects that minimize environmental impact and promote responsible production and consumption practices aligned with the principles of sustainability and circularity.

Table 1 Concepts of sustainability and circularity introduced to students during the initial brainstorming session and addressed in the diagnostic questionnaire

Sustainability	Circularity
Use of sustainable materials	Use of recycled and recyclable materials
Local production and reduced transport emissions	Transparency and traceability
Patterning/Cutting optimization	Digital fashion and virtual prototyping
Most effective garment making techniques	Waste reuse and recovery
Ethical and fair production	Design for durability
Reducing material waste	Waste reduction in pattern making and sewing
Design for longevity	Design for disassembly
Encouraging conscious consumption	Natural wear design
Transparency and traceability	Consumer education and awareness
Responsible disposal initiatives	Design for adaptability and customization
Digital fashion and virtual prototyping	Exploring new business models

3 Results

As part of the challenge, students were tasked with designing a circular skirt incorporating sustainability and circularity principles. The primary problem presented to the students was how to integrate eco-design regulations into the development of the circular skirt. While all students completed the assignment, only 45% reported applying sustainability and circularity criteria during the design phase.

Regarding the diagnostic questionnaire, Figs. 1, 2, 3 and 4 present the results for the questions related to the understanding of the concepts of sustainability and circularity, as outlined in Table 1. Figures 1 and 2 show the percentage of students who identified the different concepts of sustainability and circularity, respectively, as an associated practice. Figures 3 and 4 display the results from the questions in which students were asked to rank the importance of different concepts related to sustainability and circularity, respectively, practices on a scale from 1 to 7 (with 1 being the least important and 7 the most important).

The choice of the circular skirt added an additional technical challenge to the reflective exercise, particularly regarding pattern making, cutting, and construction. When asked to indicate the design stages in which they considered sustainability concepts, students identified the selection of materials (85%) and trims/accessories (75%) as the most relevant stages (Table 2). The results also revealed that 39% considered these principles during the ideation phase, 31% during modeling, and 27% during the manufacturing process. Overall, the importance attributed to circularity was notably lower than that given to sustainability.

When asked to rate the importance of different design stages, students identified material selection as the critical phase of the design process—that is, the stage

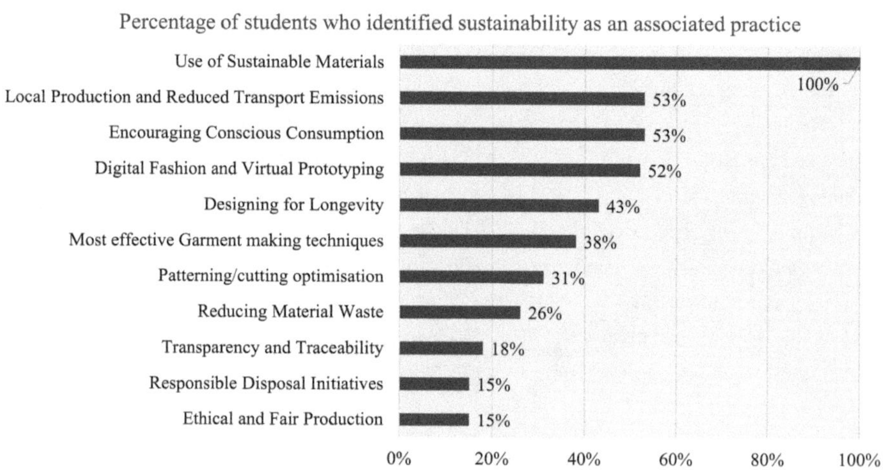

Fig. 1 Percentage of students who identified the different concepts of sustainability as an associated practice

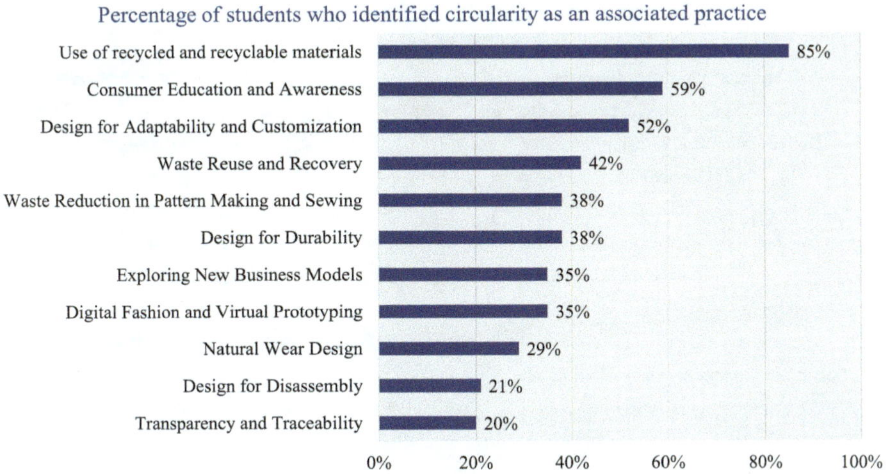

Fig. 2 Percentage of students who identified the different concepts of circularity as an associated practice

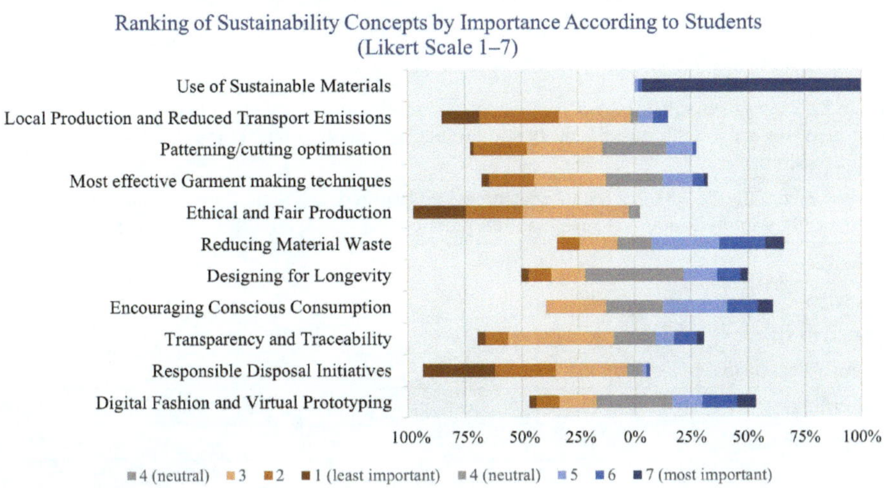

Fig. 3 Distribution of student responses (%) regarding the importance of various sustainability concepts. Responses are based on a 7-point Likert scale, where 1 indicates "least important," 4 is "neutral," and 7 indicates "most important." The length of the bars represents the percentage of students who selected each rating

they consider to have the greatest impact and importance for integrating sustainable practices—while the initial design stages, such as ideation and modeling, were considered less relevant. These results suggest a limited awareness of the potential environmental impact of decisions made during the early stages of design.

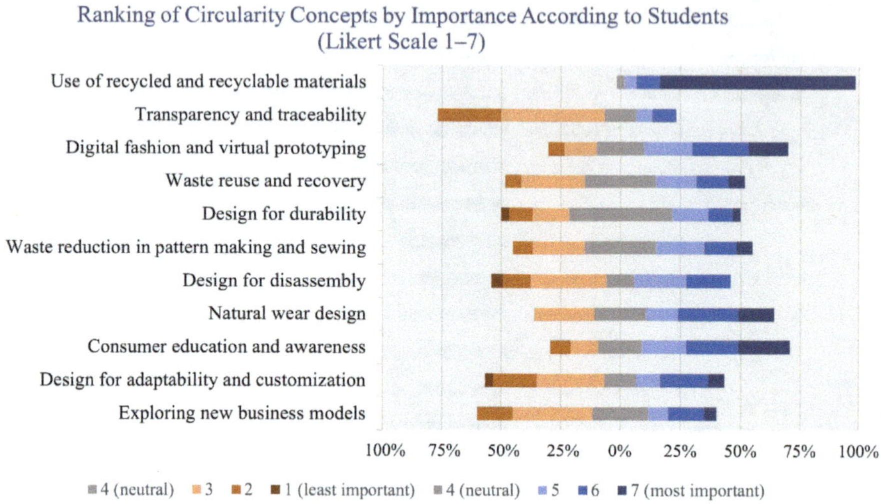

Fig. 4 Distribution of student responses (%) regarding the importance of various circularity concepts. Responses are based on a 7-point Likert scale, where 1 indicates "least important," 4 is "neutral," and 7 indicates "most important." The length of the bars represents the percentage of students who selected each rating

Table 2 Stages of the PBL methodology applied in this study and the percentage of students who identified, for each stage, sustainability and circularity practices during the development of their circular skirt projects

Stages used with the PBL methodology in this study	% of students who identified sustainability practices (%)	% of students who identified circularity practices (%)
Ideation	39	11
Sketches	42	6
Materials (fabrics)	85	26
Trims/Accessories	75	35
Patterning	31	16
Garment making	27	12

Regarding sustainable materials, the vast majority of students were unfamiliar with biomaterials or specific organic materials, such as organic wool, organic cotton, and organic linen. Regarding circularity practices, almost all students were familiar with upcycling techniques, while less than half recognized zero-waste methods, and only a small number were familiar with repair techniques.

An additional component of the project required students to present three scientific articles related to the sustainability and circularity strategies applied in their designs. However, only three students were able to identify and present articles relevant to their projects, highlighting a gap in research skills.

In the focus group discussions, only ten students were familiar with the concept of a digital product passport and its relevance in communicating product information throughout the value chain and to consumers. Although they acknowledged its importance, they lacked clarity regarding the specific data such a tool should include. Furthermore, about half of the students viewed digital tools as important for advancing sustainability, primarily in relation to consumer communication. Concepts such as traceability and transparency were not widely recognized as significant by most participants. Brainstorming sessions conducted during project development suggested that while students generated numerous ideas reflecting an understanding of sustainable design, their contributions indicated a more limited familiarity with the concept of circularity. Their perceptions of sustainable design emphasized the use of natural and biodegradable fibers, ethical production, environmental awareness, and the importance of minimizing waste and resource consumption throughout a product's life cycle. However, some students expressed a lack of motivation to incorporate sustainable solutions during the product ideation phase, noting that sustainable practices often constrained their creative choices when compared to conventional design approaches. This may be attributed to the lack of necessary resources to support students in experimenting with sustainable materials and understanding the full impact of their design decisions.

Although the project focused on the design of a circular skirt incorporating principles of sustainability and circularity, moments of discussion and reflection revealed that some students recognize the potential of design to change current consumption models.

Table 3 presents selected examples of the circular skirt projects developed by the students, including information on the sustainable practices adopted and the materials used in each case.

4 Conclusions

Universities play a crucial role in responding to the new challenges of teaching sustainability and circularity, enabling both themselves and design students to better align themselves with the European Union's guidelines for eco-design, ethical standards, social responsibility and the demands of market competitiveness.

While most students in this study demonstrated a basic understanding of sustainability concepts, their knowledge of circularity was more limited, particularly regarding zero-waste techniques and repair methods, highlighting the need for further exploration of these topics.

Only 45% of students reported applying sustainability and circularity criteria during the project development phase, indicating that, despite awareness, effectively integrating these practices remains a challenge.

While there is a growing focus on sustainability and circularity in textile and fashion design, it is equally important to continue researching effective methods to

Table 3 Examples of student projects, including the sustainable practices applied and the materials (fabrics) used in the development of the circular skirt

Project description	Sustainable practices	Materials (Fabric)	SDG's contribution
Skirt that transforms into a short coat	Product functionality with a view to increasing its life cycle Selection of sustainable materials (fabrics and trims)	Smooth, natural textured coating using B.Cork™ technology Organic cotton	4—Quality Education 6—Clean Water and Sanitation 12—Responsible Consumption and Production 13—Climate Action 14—Life Below Water 15—Life on Land 17—Partnership for the Goals
Two overlapping skirts using two sustainability criteria	Upcycling using fishing nets combined with zero waste pattern	Fishing nets	
Skirt with decorative flowers made from the waste fabric from the cut	Selection of sustainable materials Waste reduction	Recycled polyester Reused fabrics	
Skirt with decorative effects using natural dye	Selection of sustainable materials Natural dyeing	Linen	
High low skirt with ruffles	Selection of sustainable materials	Biosynthetic spider silk	
Skirt inspired by Antoni Gaudi's print from Casa Batlló	Local and ethical production Upcycling	Reused fabrics	
Long skirt in various colors	Upcycling	Reused fabrics	

teach students how to apply sustainable practices both in the conception and materialization of their projects. There is a need to improve students' skills in understanding and implementing sustainability and circularity concepts throughout the value chain.

The PBL approach facilitates the application of theoretical knowledge in a practical context, promoting a deeper understanding of the principles of sustainable fashion, circularity and strategies for tackling environmental challenges in the fashion industry.

Acknowledgements The authors are very grateful for the support granted by the Research Unit of Fiber Materials and Environmental Technologies (FibEnTech-UBI), through the Projects references UID/00195/2025 (https://doi.org/10.54499/UID/00195/2025), UID/PRR/00195/2025 (https://doi.org/10.54499/UID/PRR/00195/2025) and UID/PRR2/00195/2025 (https://doi.org/10.54499/UID/PRR2/00195/2025), funded by FCT—Fundação para a Ciência e a Tecnologia, IP/MECI.

References

1. Ellen MacArthur Foundation: A new textiles economy: redesigning fashion's future. Ellen MacArthur Foundation, Cowes, UK (2017). https://www.ellenmacarthurfoundation.org/a-new-textiles-economy, Last accessed 12 Sept 2024
2. United Nations Framework Convention on Climate Change (UNFCCC): UN helps fashion industry shift to low carbon. Accessed from: https://unfccc.int/news/un-helps-fashion-industry-shift-to-low-carbon, Last accessed 20 Mar 2025
3. Niinimäki K, Peters G, Dahlbo H, Perry P, Rissanen T, Gwilt A (2020) The environmental price of fast fashion. Nat Rev Earth Environ 1(4):189–200. https://doi.org/10.1038/s43017-020-0039-9
4. Sarokin SN, Bocken NMP (2024) Pursuing profitability in slow fashion: exploring brands' profit contributors. J Clean Prod 444:141237. https://doi.org/10.1016/j.jclepro.2024.141237
5. Leal Filho W, Dinis MAP, Liakh O et al (2024) Reducing the carbon footprint of the textile sector: an overview of impacts and solutions. Textile Res J **94**(15–16):1798–1814. https://doi.org/10.1177/00405175241236971
6. Sung K (2023) Understanding upcycling and circular economy and their interrelationships through literature review for design education. In: Proceedings of the design society, vol 3, pp 3721–3730. https://doi.org/10.1017/pds.2023.373
7. Fletcher K, Tham M (2019) Earth logic fashion action research plan. Kate Fletcher and Mathilda Tham, London. https://katefletcher.com/wp-content/uploads/2019/10/Earth-Logic-plan-FINAL.pdf, Last accessed 20 Mar 2025
8. Casciani D, Wang W (2023) Unpacking strategies for E-textile design for disassembly. In: Heinzel T, Dumitrescu D, Tomico O, Robertson S (eds) Proceedings of textile intersections conference 2023. London, United Kingdom. https://doi.org/10.21606/TI-2023/117
9. Council of the EU: "Green transition: Council gives its final approval to the ecodesign regulation"—Council of the EU, Press release, 27 May 2024. https://www.consilium.europa.eu/en/press/press-releases/2024/05/27/green-transition-council-gives-its-final-approval-to-the-ecodesign-regulation/, Last accessed 12 Sept 2024
10. European Union: Regulation (EU) 2024/1781 of the European Parliament and of the Council of 13 March 2024 establishing a framework for the setting of ecodesign requirements for sustainable products. Official J European Union L 2024/1781 1–89. https://eur-lex.europa.eu/eli/reg/2024/1781/oj
11. Nguyen LTV, Cleveland D, Nguyen CTM, Joyce C (2024) Problem-based learning and the integration of sustainable development goals. J Work-Appl Manag 16(2):218–234. https://doi.org/10.1108/JWAM-12-2023-0142
12. Choi KH (2018) Eco-tech fashion project: collaborative design process using problem-based learning. Int J Fashion Design Technol Educ 12(1):105–117. https://doi.org/10.1080/17543266.2018.1516808
13. Hmelo-Silver CE (2004) Problem-based learning: what and how do students learn? Educ Psychol Rev 16:235–266. https://doi.org/10.1023/B:EDPR.0000034022.16470.f3
14. Gam HJ, Banning J (2011) Addressing sustainable apparel design challenges with problem-based learning. Cloth Text Res J 29(3):202–215. https://doi.org/10.1177/0887302X11414874
15. Ma JJ (2023) Development of education for sustainable fashion design using a challenge-based learning approach. Int J Fashion Design Technol Educ 16(2):164–174. https://doi.org/10.1080/17543266.2022.2137249
16. Fashion for Change EU: Supporting Sustainable Fashion Designers, Start-Ups and SMEs in the Circular Economy (2021). https://www.fashionforchange.eu/wp-content/uploads/2021/11/F4C-Report-Supporting-Sustainable-Fashion-Start-Ups_2021.pdf, Last accessed 20 Mar 2025
17. Textile Exchange: 10th Edition of Textile Exchange's Annual Materials Market Report (2023). https://textileexchange.org/knowledge-center/reports/materials-market-report-2023/, Last accessed 08 Sept 2024

18. Bicho M, Pereira M, Barata TR, Belino N, Miguel R (2024) Sustainable fashion product development: the importance of skills. In: Cunha J et al (eds) New materials and processes in the academy and industry context, advances in fashion and design research II. Springer, Cham, pp 287–301. https://doi.org/10.1007/978-3-031-43937-7_26
19. Ley K, van Mazijk R, Boger S, Watten D, Pardo CM, Zuo C (2020) Financing the Transformation in the Fashion Industry: Unlocking Investment to Scale Innovation. Fashion for Good. https://fashionforgood.com/wp-content/uploads/2020/01/FinancingTheTransformation_Report_FINAL_Digital-1.pdf, Last accessed 12 Sept 2024
20. Lyons P, Bandura R (2025) A reciprocal–reflective approach to learning: concept and model. Euro J Training Dev 49(1/2):210–223. https://doi.org/10.1108/EJTD-09-2023-0133
21. Zhang R (2021) Design of an authentic project-based learning course using project management approach. Int J Fashion Design Technol Educ 15(1):57–66. https://doi.org/10.1080/17543266.2021.1992799
22. Bocken NMP, de Pauw I, Bakker C, van der Grinten B (2016) Product design and business model strategies for a circular economy. J Ind Prod Eng 33(5):308–320. https://doi.org/10.1080/21681015.2016.1172124

mHealth Applications and the Challenges Faced by the Elderly: A Literature Review

M. Vilas Boas, R. Gonçalves, N. Martins, and R. Simoes

Abstract The rise of mHealth solutions offers significant opportunities to improve healthcare access and management for older adults. However, age-related cognitive, physical, and perceptual barriers limit the usability and acceptance of these technologies among the elderly population. This article presents a literature review focused on understanding the key challenges faced by the elderly when interacting with mHealth applications and identifying the main design principles that can enhance usability, accessibility, and user satisfaction. The review synthesizes evidence from studies on human factors, gerontology, and health technology design, highlighting the impact of cognitive decline, reduced motor skills, visual impairments, and low digital literacy on mobile health app usage. It also addresses motivational barriers, determinants of technology acceptance, and how intuitive design can be used to promote engagement and autonomy. The findings highlight the importance of adopting user-centered and inclusive design practices that accommodate the diverse needs of older users, reducing complexity and improving navigation, readability, and interaction. This analysis provides a valuable foundation for designers, developers, and researchers involved in the development of mHealth applications targeted at the elderly, supporting solutions that foster better treatment adherence, patient empowerment, and quality of life.

Keywords mHealth · Interface design · Senior population · Elderly · Mobile health

M. Vilas Boas · N. Martins
Media and Culture, School of Design, Research Institute for Design, Polytechnic of Cávado and Ave, Barcelos, Portugal
e-mail: a29959@alunos.ipca.pt

N. Martins
e-mail: nmartins@ipca.pt

R. Gonçalves (✉) · R. Simoes
2Ai – School of Technology, IPCA - Polytechnic Institute of Cávado and Ave, Barcelos, Portugal
e-mail: rtgoncalves@ipca.pt

R. Simoes
e-mail: rsimoes@ipca.pt

D. Raposo et al. (eds.), *Strategic Design for Sustainable Futures*, Springer Series in Design and Innovation 65, https://doi.org/10.1007/978-3-032-33282-0_7

1 Introduction

Mobile health (mHealth) has been transforming healthcare by offering innovative solutions for various chronic conditions, including lymphedema. With advancements in mobile technologies, there is an opportunity to develop digital tools aimed at facilitating treatment adherence, promoting patient autonomy, and improving quality of life, particularly among the elderly population. The challenges in managing lymphedema highlight the need for innovative solutions, such as mHealth technologies. This type of mobile advancements provides tools that can improve self-management and provide remote support, enhancing the adherence to treatment regimens. These technologies can empower patients to take charge of their condition while providing healthcare professionals with critical data to personalize care [1]. A mobile application connected to a lymphedema treatment device could play a crucial role in facilitating remote care and improving patient management. The app can allow healthcare providers to monitor treatment progress from a distance, enabling consistent oversight without the need for frequent in-person visits. This capability is particularly valuable for individuals with chronic conditions like lymphedema, where long-term management is essential, and regular travel to healthcare facilities can be burdensome. The convenience of remote treatment also empowers patients by giving them greater autonomy over their care [1].

This article is part of a research project in collaboration with CeNTI-Center for Nanotechnology and Advanced Materials, which aims to develop an intuitive mHealth application for managing therapeutic sleeves used in lymphedema treatment. Adherence to this type of therapy is a critical factor for treatment success. However, many patients, particularly older adults, face difficulties in using complex technologies, compromising the effectiveness of compression therapy.

In this context, this study focuses on a literature review of the usability and accessibility of health technologies targeted at older adults. The goal is to support the creation of a simple, intuitive, and tailored interface for this audience. Analyzing the challenges that the elderly face when using mHealth applications and the design recommendations to overcome these barriers will be central to the application's proposal to ensure that the solution is both effective and accessible. At the end of the project, the application is expected not only to improve lymphedema treatment outcomes but also to enhance patient autonomy and healthcare efficiency.

2 mHealth

The aging population is a growing trend that poses significant challenges to healthcare systems. In response to this situation, technological tools have played a crucial role in care delivery, especially in eHealth and mHealth. eHealth refers to health services that use information and communication technology (ICT), such as computers and mobile phones, to provide health services and information, while mHealth focuses more specifically on the use of smart or portable devices for these purposes [2]. These technological solutions not only promote efficiency in care delivery but also enhance patient engagement in their treatment, which is essential for vulnerable populations such as older adults.

As the population ages, mHealth tools are becoming increasingly common. These technologies help older adults live more independently, allowing them to maintain their health without frequent medical appointments, which also reduces the risk of cross-contamination in in-person encounters [3]. They also provide convenience for seniors, often replacing regular visits to doctors while keeping them connected to their healthcare remotely.

mHealth has been highlighted as a promising resource to encourage patients to actively participate in their healthcare. One of the reasons is that most people own and regularly use a mobile phone, making these applications ideal for providing health-related information and support. Moreover, it can be a powerful behavior change tool, promoting health prevention and self-management due to the ubiquity and advanced computational capabilities of mobile devices [4].

However, the success of these solutions depends on their acceptance and ease of use by the elderly. Many devices fail because developers do not prioritize usability, resulting in complex interfaces that interfere with the interaction and lead to errors, such as incorrect option selection [5, 6]. As mHealth technology evolves, it is essential to develop effective design mechanisms collaboratively, involving end users in the design process. This ensures applications are easy to use and perceived as useful. In fact, many current mHealth interventions are based on pre-existing structures within the healthcare system, potentially limiting their effectiveness compared to those designed specifically for users' needs [4].

Beyond helping patients monitor their health, this technology also has the potential to ease the burden on healthcare systems. For instance, portable devices that monitor the health of older adults with chronic diseases have proven effective in improving treatment adherence, aiding healthcare professionals by allowing for daily treatment adjustments [7]. Telemedicine, or telehealth, is an eHealth solution that has been particularly useful in elderly care. It involves the use of technology to exchange medical information electronically across different locations, improving patients' health without the need for physical mobility [8].

Originally, telemedicine referred to interactions between doctors and patients via telecommunications. Today, telehealth is a broader and more suitable term, encompassing interactions among different healthcare professionals [9]. The adoption of telehealth can improve patient experiences, reduce healthcare costs, and enhance healthcare professionals' work satisfaction [8].

For elderly people living with chronic diseases, telehealth offers the advantage of quicker and more autonomous access to health information. Remote monitoring technology has been associated with a reduction in unnecessary hospitalizations, thereby easing the pressure on the healthcare system and enabling doctors to adjust treatments more efficiently [7, 10]. Various studies also indicate that mHealth applications, by providing immediate and continuous access to health information, can significantly improve clinical outcomes and reduce the risk of complications [11].

Self-management of healthcare is one of the areas where mHealth solutions can have the greatest impact, as long as they are well-accepted by older adults.

Nevertheless, despite this promising outlook, mHealth and eHealth usage also present challenges. A study by Agnihothri et al. [12] identified several pros and cons of using mHealth applications. Among the pros are convenience, personalized care, increased accuracy in health data, and effective information sharing between caregivers and patients. MHealth applications can also enhance treatment adherence through reminders and messages and facilitate prompt interventions by nurses or automated systems, potentially reducing direct and indirect costs. Additionally, mHealth contributes to the safety of both older adults and healthcare professionals, accelerates service delivery, reduces associated costs, and lowers risks of illness and death [13].

On the other hand, the cons include the need for training for patients that are unfamiliar with technology with the potential inaccuracy of patient-reported data, and concerns about device reliability, as they are not always certified medical devices. Moreover, issues related to privacy and data protection, lack of integration with electronic health records (EHR), and additional time required for healthcare professionals to monitor patient-uploaded information are also significant challenges [12].

The success of mHealth and eHealth tools largely depends on their acceptance by the target audience, especially elderly people. Intuitive design and ease of use must be priorities to ensure these technologies truly benefit this population. When implemented effectively, mHealth and eHealth tools can significantly improve the management of health conditions and the quality of life for older adults while helping control costs and reduce the strain on healthcare systems [12].

According to Liu et al. [14], mHealth application designs have been evaluated in two dimensions: end-user perceptions and health-related outcomes. Examples of end-user perceptions include usability, viability, acceptance, adherence, compliance, engagement, satisfaction, adoption, and usage, as well as feedback during the use, such as facilitators, barriers, and suggestions for future improvements. Health-related outcomes, on the other hand, include behavioral changes, physical and cognitive functions, psychosocial and emotional well-being, quality of life, and health literacy.

3 Challenges for the Senior Population

The aging population and the fast advancement of technology present numerous challenges for interacting with digital technologies. The declines in cognitive, visual, and motor skills directly affect the usability of digital interfaces. Additionally, factors like digital literacy and cultural barriers further complicate the adoption of new technologies. This scenario demands a careful approach to the application design in order to optimize user experience and promote greater digital inclusion.

The elderly face various health challenges related to aging. However, they now enjoy a longer average life expectancy thanks to improved access to healthcare compared to previous generations. Technological advances have played a crucial role in promoting the well-being of seniors by facilitating communication and access to healthcare services [3].

Martins et al. [15] argue that senior users are complex individuals who, despite sharing common traits within their age group, also exhibit individual characteristics influenced by demographic and psychographic factors. Demographic aspects include race, ethnicity, education level, health status, and family structures, while psychographic traits, such as personality, motivation, attitude, and interests, play a significant role in the adoption of new technologies.

According to Wahl and Tesch-Römer (as cited in Chun and Patterson [16]), older adults naturally experience differences in physical and mental abilities compared to younger adults. These limitations can affect how seniors use interfaces in telemedicine systems. Moreover, when a homepage lacks clear guidance, seniors face difficulties in finding what they need, often taking longer to navigate or feeling confused about where to click for the correct content.

Fear of making mistakes and the lack of self-confidence are significant obstacles to adopting digital technologies for older adults. This demographic often shows less willingness to learn and fears damaging the devices [17].

Furthermore, Chun and Patterson [16] highlight an interesting observation, that the seniors often blame themselves when they cannot complete tasks rather than criticizing the system's interface. This may be because they lack confidence in using the internet especially for telemedicine systems they don't use regularly, even if they are familiar with general internet navigation. It's also worth highlighting that one of the main challenges in designing telemedicine interfaces for older people lies in organizing and structuring health information. Understanding the user needs is essential for the success of an information system, including mobile applications [6].

According to Morey et al. [5], a lack of attention to these issues has led to usability problems for mHealth applications. To truly serve the elderly population, such applications must be tailored to their needs and characteristics. Particularly for older people, mHealth applications can be challenging due to complex menus, small text fonts, and other usability issues. Seniors struggle with features such as screen size, device dimensions, buttons, colors, and backlighting, as well as functionalities like memory supports, operational complexity, layout, and shortcuts. However, increased use of communication applications and other mobile apps could improve this situation [18].

To better understand the requirements for senior-friendly mHealth applications, Tajudeen et al. [6] conducted semi-structured interviews with 74 seniors aged between 60 and 84 (average age is 68). The participants owned mobile phones and had experience using at least two or three mobile applications. During the interviews, images of mobile application screens were shown to identify preferences for text, color, menu layout, background and other elements.

To further assess seniors' familiarity with mobile applications, researchers identified the most frequently used apps. It was observed that most seniors could recognize and use common application functions, such as logging in/out, accessing the home screen, clicking hyperlinks, playing videos, listening to audio, entering information, and using common buttons (e.g., save, send, cancel, delete). The authors also noted that seniors navigate efficiently through mobile applications they find easy to use and engaging. One user reported frustration when there are too many elements in the interface, making it difficult to locate desired features, leading them to avoid such apps if it took too long to find what they needed.

Subsequently, Tajudeen et al. [6] sought to better understand the concerns of the senior population regarding the use of mobile applications. Through this analysis, they identified several challenges faced by the elderly when using these applications. These challenges can be grouped into four categories: the utility of the application, complexity, technical knowledge required, and necessity.

Firstly, application utility was a notable concern. Despite having experience with various applications, many seniors still found them not particularly useful or relevant to their daily routines. For them, the presence or absence of mobile applications does not significantly alter their activities, as their daily tasks are often limited [6].

The second challenge was complexity. Many seniors expressed frustration with interfaces featuring excessive information, such as text, images, videos, links, and unlabeled icons, as well as tasks requiring multiple steps. They often lose focus and concentration when attempting to perform tasks in such applications. The authors clarified that situations perceived as simple by younger users can seem complex to seniors, leading to frustration and hesitation, especially if assistance is unavailable [6].

Additionally, seniors face difficulties with the required technical knowledge. Many participants mentioned that they did not know how to install applications on their devices. Assistance may be needed to install, register and configure apps for the first time. Those without family or friends to provide this support might avoid using apps overall.

Lastly, necessity emerged as a challenge. While there is a wide range of mobile applications available for social interaction and quality-of-life improvement, seniors demonstrated a preference for apps that facilitate communication and leisure, such as Facebook and WhatsApp. Seniors feel that these apps are enough for their needs, as using many different applications can be overwhelming and stressful [6].

3.1 Barriers in Aging

In general, elderly people face various health issues related to aging. According to Ramdowar et al. [3], age-related conditions, such as mental disorders, can lead to depression or anxiety. The same authors identify three main barriers associated with aging that limit the use of mHealth applications by seniors: cognitive, perceptual, and physical barriers. Since the adoption of mobile applications by seniors remains low, limitations related to their perception, cognition, and physical abilities are often overlooked in the design of mHealth interfaces.

3.1.1 Cognitive Barriers

Cognition, which is closely related to memory, tends to be affected by aging, leading to increased time needed to process information and difficulties in recalling it. Individuals facing these challenges experience greater difficulty using mHealth applications. In this context, studies show that older adults prefer interfaces with fewer steps and more simplified navigation. Additionally, consistent use of terminology is recommended to reduce confusion caused by unfamiliar words, avoiding memory overload with complex information [3]. Issues related to reduced cognitive speed and capacity are also common. As cognitive abilities decline, tasks requiring substantial cognitive effort and processing take longer to be completed by older people [17].

It is important to note that the aging process is associated with a decline in cognitive abilities, which hinders the assimilation of new information and learning. However, despite these challenges, not all types of memory are equally affected by aging. For instance, short-term memory, working memory, semantic memory, prospective (future) memory, and perceptual speed are distinct types of memory that can be impacted in different ways by aging [19].

Short-term memory is responsible for storing information that has just been received or understood. Such information tends to be quickly lost if not frequently reviewed or used. With aging, the ability to utilize this information becomes slower [20].

As for working memory, it refers to a person's ability to hold specific information while simultaneously performing another activity. During tasks that require temporarily retaining information for later use, it is recommended that, when designing an interface, older users should not rely on this type of memory. Instead, the interface should present only the information relevant to the task at hand, reducing the need to depend on working memory, which weakens with age [21].

On the other hand, semantic memory refers to the accumulation of factual knowledge acquired over a lifetime, including vocabulary, historical facts, cultural norms, language and more. Unlike other types of memory, semantic memory does not deteriorate with aging, and complete loss of stored content is rare. From a design perspective, leveraging semantic memory, particularly through culturally familiar references, can facilitate the use of interfaces by seniors [22].

Prospective memory, in turn, refers to the ability to remember to perform a task at a future time. It can be divided into two types: time based, and event based. Time based memory, such as the ability to remember to take medication at specific times, tends to decline with age. Event based memory, which involves remembering to perform a task after an event occurs, is less affected by aging [21].

Perceptual speed, related to the ability to quickly identify and compare different visual stimulation, is another cognitive ability that decreases with age. This skill is essential for various human cognitive activities, such as memorizing and entering a phone number or learning the rules of a new game, and it declines steadily as one ages [23].

Additionally, another cognitive ability affected by aging is the capacity to focus visual attention. Older people face greater challenges in concentrating on visual tasks and shifting their attention from one point to another. This dynamic visual attention, which involves the ability to quickly change focus, is slower in older adults compared to younger individuals. Thus, issues like failing to observe the screen while typing or diverting attention from what is displayed on a digital interface are common. To improve usability, it is crucial for the interface to be visually clear and avoid distracting elements that could confuse or divert the attention of older users [24, 25].

The ability to understand language is also affected by aging, although it is not entirely compromised. This skill remains intact when the terms used are part of the elderly's semantic memory, meaning the vocabulary is familiar and learned over a lifetime. However, when a term of concept is not immediately clear, older people must make a cognitive deduction. This deduction depends on efficient short-term memory, which tends to weaken with age. As a result, older individuals encounter greater difficulty interpreting language when such deductions are necessary [23].

Challenges in acquiring procedural knowledge are also evident. Procedural knowledge refers to familiarity with the steps required to complete a task. Procedures that were well understood in the past tend to be retained into old age, making them difficult to forget. This means older adults generally take longer to learn new procedures compared to younger individuals. In other words, new patterns and methods that are unfamiliar are often difficult for them to understand and learn. This issue becomes even more pronounced when they are required to memorize multiple combinations of steps and their implications [22].

Finally, there are additional challenges for seniors when learning new procedures. They may become impatient and attempt to use new applications without investing much time in understanding how they work. This suggests that this demographic might feel discouraged from acquiring new knowledge, particularly when dealing with unfamiliar systems [17].

Users with low digital literacy interact differently with digital platforms compared to those with moderate or high literacy. They tend to conduct searches in a scattered and unfocused manner, spending more time on each page. Typically, they take about eight times longer to complete a task on a digital interface than the average user. Additionally, they are more likely to get lost while navigating and are more inclined to accept the first search results, even if they are not ideal [26].

Moreover, people with low digital literacy struggle to process information, focusing on one element at a time and avoiding complex content. Due to short-term memory lapses and limited vision, they often forget actions such as using the back button or scrolling [27].

These users also find hierarchical menus challenging, leading to incorrect interpretations and trial-and-error navigation. They have difficulty using search engines, which causes spelling errors that further hinder results. Visually, they struggle to interpret detailed images or abstract icons, preferring figurative and simple elements instead [23].

Cultural barriers also present significant obstacles to older adults adopting digital technologies, often tied to digital illiteracy. Due to the quick evolution of technologies, the senior population finds it more difficult to keep up with these changes. Additionally, their lack of familiarity with computers and the internet, combined with difficulties understanding digital conventions and terminologies aggravates the issue [17].

3.1.2 Perceptual Barriers

Hearing ability is one of the skills that deteriorates with age, affecting how people interact with sounds. Furthermore, simply increasing the volume is not an effective solution, as it can amplify background noise and increase confusion. Additionally, older people struggle to distinguish similar sounds and follow higher-pitched voices, such as female voices, presenting a challenge in designing interfaces for this population [17].

Vision is another ability that is significantly affected by aging. Among the various factors associated with user interface issues, font size is the most significant. With reduced visual acuity, many older adults find it difficult to read text in the standard size used in applications, preferring larger fonts [3].

In this context, visual problems such as macular degeneration and the use of bifocal glasses affect more than 20% of seniors. As a result, these individuals face difficulties viewing digital screens and require more light to see clearly. Additionally, reduced visual precision hinders readability, making small text sizes inaccessible to older users. According to Ramdowar et al. [3], some experts recommend implementing features such as a zoom function or voice assistants to improve accessibility.

Although vision impairment is not exclusive to aging, the prevalence of ocular issues increases with age. Visual acuity, which refers to the sharpness of image perception, typically begins to decline around the age of 40. This decline is mainly due to presbyopia, which impairs the ability to focus on nearby objects, and a reduced amount of light reaching the retina, which yellows over time [25].

As a result, many older people require reading glasses and become more sensitive to intense light. Additionally, color differentiation and depth perception are also affected, particularly with short-wavelength colors like green and blue. Between the ages of 55 and 65, changes in the retina and central nervous system reduce the useful visual field and impair the ability to adapt to darkness [17].

In this regard, other visual difficulties that arise include reduced contrast sensitivity, slower visual information processing, and difficulties distinguishing similar objects. Over 10% of older adults suffer from degenerative diseases, making moving visual elements like animations inadvisable [23].

These changes in vision significantly impact screen usage, as many seniors may struggle even to initiate interactive tasks due to difficulties reading what appears on the screen. This leads to increased time spent locating icons and confusion when differentiating them, particularly when they are similar in shape, color and size. Therefore, it is crucial for digital applications with visual interfaces to account for these limitations when designing for this population [23].

As we can see, aging affects the interaction of older people with new digital technologies for various reasons. The skills required to use graphical digital interfaces-such as visual acuity, memory and motor dexterity- are among those that deteriorate most with age. This makes it harder for older people to view interfaces, build new conceptual models, and use technological devices. These difficulties reduce their confidence in handling new situations and create resistance to addressing new challenges [28].

Finally, with the increasing use of touchscreen interfaces, tactile perception has gained significant relevance in interface design for older adults. Pereira et al. [17] define haptics as "the science that studies touch and its sensations as a means of control and interaction with machines and computers". The authors argue that it is essential to understand how aging affects this ability to optimize the user experience. They highlight that aging leads to greater variability in tactile control capabilities and expands the thresholds of tactile perception, such as temperature and vibration. This suggests that older people may experience reduced sensitivity to tactile feedback, reducing the effectiveness of tactile feedback in interface design.

3.1.3 Physical Barriers

The loss of motor skills, a common physical barrier, significantly impacts older adults' ability to use mHealth applications. With age, movements slow down, and muscles and joints stiffen, making it harder to interact with small buttons or links. Consequently, seniors prefer having all icons or options visible on a single screen to avoid the need for scrolling. The use of large icons is therefore recommended to improve touch accuracy [3].

Aging affects motor skills, beginning with increased movement time due to diminished muscle strength and endurance. Older adults are typically 1.5–2 times slower than younger individuals in executing movements, with delays more evident in decision-making than in the physical execution of movements. This increased response time can complicate interactions with digital interfaces. For instance, slow clicks may be misinterpreted by the system, causing frustration [17].

Movement precision also declines with age. As a result, older adults may struggle to adjust movement trajectories or apply appropriate force, reducing the accuracy of their actions. This issue is particularly critical when interacting with digital interfaces, where precise clicking or tapping is essential [17].

Finally, movement coordination deteriorates with age, making it harder for seniors to synchronize multiple movements. For example, picking up an object requires moving the arm toward it and opening the fingers to grasp it, a task that becomes increasingly difficult for older adults [17].

4 Conclusion

The development of health technologies for the elderly, such as mHealth applications, holds significant potential to improve the quality of life for this demographic, particularly in the context of chronic conditions like lymphedema. This article explored the main challenges that seniors face when using digital technologies, addressing issues related to vision, hearing, motor skills, and cognitive limitations (such as memory and attention). The need to simplify interfaces to facilitate use and promote greater user autonomy was a recurring theme throughout the investigation.

The proposed mHealth application aims to address these specific challenges, promoting adherence to lymphedema treatment and encouraging independence among elderly patients. By focusing on user experience and adapting functionalities to meet the limitations and needs of this population, the application can offer continuous and personalized support, simplifying the process of monitoring and controlling therapeutic sleeves.

Despite these advancements, the field of health technologies for older adults remains an emerging area with room for future investigations that delve deeper into the development of innovative design practices. While this article focused on identifying the challenges faced by seniors in using digital technologies and proposed a conceptual mHealth application to support adherence to lymphedema treatment, the development of specific design recommendations was not explored in detail. However, creating clear and tailored guidelines for designing interfaces for this population is crucial to ensuring technological solutions are genuinely effective.

Future research should prioritize the formulation of design recommendations that address the challenges discussed throughout this article, such as visual, motor and cognitive limitations, as well as the need for intuitive and simplified interfaces. Investigations in this area could provide valuable guidelines for developing more inclusive and adaptable mHealth applications, fostering positive user experience and supporting the autonomy of older adults in managing their health.

Acknowledgements This work featured in this paper was supported by a Research Grant (HfPT—PPS3-IPCA-ESD-2AI-BL-1) awarded by the Polytechnic Institute of Cávado and Ave (IPCA). The grant is part of the Health from Portugal (HfPT) Innovation Pact [C630926586-00465198], co-financed by the Recovery and Resilience Plan (RRP) and by the European funds NextGeneration EU (https://recuperarportugal.gov.pt/), through the Incentive System "Agendas for Business Innovation". Foundation for Science and Technology (Lisbon), through projects UIDB/05549/2020, UIDP/05549/

References

1. Mangion A, Ivasic B, Piller N (2024) The utilization of e-health in lymphedema care: a narrative review. Telemed e-Health. 30:331–340. https://doi.org/10.1089/tmj.2023.0122
2. Moss RJ, Süle A, Kohl S (2019) EHealth and mHealth. Eur J Hosp Pharm 26:57–58. https://doi.org/10.1136/ejhpharm-2018-001819

3. Ramdowar H, Khedo KK, Chooramun N (2023) A comprehensive review of mobile user interfaces in mHealth applications for elderly and the related ageing barriers
4. Schnall R, Rojas M, Bakken S, Brown W, Carballo-Dieguez A, Carry M, Gelaude D, Mosley JP, Travers J (2016) A user-centered model for designing consumer mobile health (mHealth) applications (apps). J Biomed Inform 60:243–251. https://doi.org/10.1016/j.jbi.2016.02.002
5. Morey SA, Stuck RE, Chong AW, Barg-Walkow LH, Mitzner TL, Rogers WA (2019) Mobile health apps: improving usability for older adult users. Ergon Design 27:4–13. https://doi.org/10.1177/1064804619840731
6. Tajudeen FP, Bahar N, Tan MP, Peer Mustafa MB, Saedon NI, Jesudass J (2022) Understanding user requirements for a senior-friendly mobile health application. Geriatrics (Switzerland) 7. https://doi.org/10.3390/geriatrics7050110
7. Loera JA (2008) Generational differences in acceptance of technology. Telemed e-Health. 14:1087–1090. https://doi.org/10.1089/tmj.2008.0028
8. Tuckson RV, Edmunds M, Hodgkins ML (2017) Telehealth. N Engl J Med 377:1585–1592. https://doi.org/10.1056/NEJMsr1503323
9. Wootton R (2001) Telemedicine. BMJ 323:557–560. https://doi.org/10.1136/bmj.323.7312.557
10. Wildenbos GA, Jaspers MWM, Schijven MP, Dusseljee-Peute LW (2019) Mobile health for older adult patients: using an aging barriers framework to classify usability problems. Int J Med Inform 124:68–77. https://doi.org/10.1016/j.ijmedinf.2019.01.006
11. Barra DCC, Paim SMS, dal Sasso GTM, Colla GW (2017) Methods for developing mobile apps in health: an integrative review of the literature
12. Agnihothri S, Cui L, Delasay M, Rajan B (2020) The value of mHealth for managing chronic conditions. Health Care Manag Sci 23:185–202. https://doi.org/10.1007/s10729-018-9458-2
13. Abbaspur-Behbahani S, Monaghesh E, Hajizadeh A, Fehresti S (2022) Application of mobile health to support the elderly during the COVID-19 outbreak: a systematic review
14. Liu N, Yin J, Tan SSL, Ngiam KY, Teo HH (2021) Mobile health applications for older adults: a systematic review of interface and persuasive feature design
15. Martins N, Ralha S, Simoes R (2022) ICT4Silver: design guidelines for the development of digital interfaces for elderly users. In: Lecture notes in mechanical engineering. Springer Science and Business Media Deutschland GmbH, pp 178–188
16. Chun YJ, Patterson PE (2012) A usability gap between older adults and younger adults on interface design of an Internet-based telemedicine system. In: Work, pp 349–352
17. Pereira L, Brandão D, Martins N (2021) Ageing related human factors to be addressed in the design of visual interfaces of digital applications developed for seniors: a literature review. In: Springer series in design and innovation. Springer Nature, pp 65–80
18. Wong CY, Ibrahim R, Mansor E (2018) usability and design issues of smartphone user interface and mobile apps for older adults. In: 5th international conference, i-USEr 2018, Puchong, Malaysia, August 28–30, 2018, Proceedings. Presented at the August
19. Anderson ND, Murphy KJ, Troyer AK (2024) How memory typically changes with age. In: Living with mild cognitive impairment. Oxford University Press, New York, pp 8–15
20. Vallar G (2017) Short-term memory. In: The curated reference collection in neuroscience and biobehavioral psychology. Elsevier, pp 367–381
21. Light LL (2024) Memory and aging. In: The Oxford handbook of human memory, two volume pack. Oxford University Press, pp 1600–1644
22. Brickman AM, Stern Y (2009) Aging and memory in humans. In: Encyclopedia of neuroscience. Academic, pp 175–180
23. Pak R, McLaughlin A (2010) Designing displays for older adults. CRC Press
24. Ren Y, Xu Z, Zhang Y, Li J, Wu J, Yang W (2019) The effect of aging on attentional networks. In: Proceedings of 2019 ieee international conference on mechatronics and automation, ICMA 2019. Institute of Electrical and Electronics Engineers Inc., pp 543–548
25. Madden DJ (2007) Aging and visual attention. Curr Dir Psychol Sci 16:70–74. https://doi.org/10.1111/J.1467-8721.2007.00478.X/ASSET/69EEFD07-E153-4EBF-9B11-60B7066B715F/ASSETS/IMAGES/LARGE/10.1111_J.1467-8721.2007.00478.X-FIG2.JPG

26. Alton NT (2020) The digital divide: how low-literate freshman search for information. In: Advances in intelligent systems and computing. Springer, Cham, pp 380–386
27. Guimarães L, Martins N, Pereira L, Penedos-Santiago E, Brandão D (2022) Interface design guidelines for low literature users: a literature review. In: ACM international conference proceeding series. Association for Computing Machinery, pp 29–35
28. Xu L, Lam YY, Ng DTK, Peng P, Suen WL, Lee CSY, Chu SKW (2024) Digital literacy education for older adults: a scoping review. In: Proceedings of the association for information science and technology. Wiley, pp 691–696

Strategic Design and Systemic Governance for Sustainable Development

Design as a Considered Driver for Organizational Growth: Transforming Creative Practice into Value Creation Insight

Jorge Brandão Pereira

Abstract Design has evolved into a strategic and integrative discipline that plays a pivotal role in shaping innovation, organizational growth, and territorial development. Positioned at the intersection of culture, creativity, and economic transformation, design contributes to competitiveness, identity-making, and systemic change across diverse contexts. This article explores the role of design in the creative economy through a multidisciplinary lens, analyzing its potential to generate value beyond aesthetics by influencing management practices, public policy, and regional ecosystems. Drawing on a qualitative, comparative case study methodology, the research examines four European initiatives—Designregio Kortrijk, PROUD, Luxinnovation, and Cité du Design—that strategically embed design in economic and social frameworks. Each case highlights how design can foster sustainable innovation, collaborative governance, and inclusive growth, while also revealing challenges such as institutional fragmentation, engagement sustainability, and the risk of model trivialization. The analysis is grounded in updated theoretical and policy frameworks that link design to broader transitions toward knowledge-based and creative societies. By aligning empirical insights with strategic reflection, the study contributes to understanding design as a mediator of transformation—both symbolic and practical—capable of enabling resilience and value creation in real-world urban, organizational, and territorial systems.

Keywords Design culture · Innovation ecosystems · Creative economy · Strategic design

J. Brandão Pereira (✉)
Institute for Research in Design, Media and Culture (ID+), 4750-810 Barcelos, Portugal
e-mail: jmpereira@ipca.pt

Design School, Polytechnic University of Cávado and Ave, 4750-810 Barcelos, Portugal

D. Raposo et al. (eds.), *Strategic Design for Sustainable Futures*, Springer Series in Design and Innovation 65, https://doi.org/10.1007/978-3-032-33282-0_8

1 Introduction

Design's role in shaping innovation and development has gained renewed interest amid the transition toward knowledge-based and creative economies [1, 1]. No longer confined to aesthetics or product development, design is now recognized as a strategic function embedded in governance, entrepreneurship, and community engagement. The increasing complexity of societal challenges, such as urban regeneration, digital transformation, environmental sustainability, has underscored the need for integrative approaches where design operates as both method and mindset.

Historically regarded as a predominantly practice-oriented discipline, this view is rooted in the emergence of design as a theoretical field during the modern movement of the twentieth century [3]. Definitions of design range from specific to broad, addressing ideas, processes, outcomes, and overall benefits. As Buchanan notes, "No single definition of design, or branches of professionalized practice such as industrial or graphic design, adequately covers the diversity of ideas and methods gathered under the label" [4]. The evolution of design methodologies is closely intertwined with developments in knowledge production and cultural practices originating in adjacent fields.

Design increasingly engages with issues beyond its traditional domain; it is no longer confined to conventional roles. Positioned as a driver of innovation, design seeks to transform realities through creativity. Its integrative nature fosters interaction across diverse sectors and among multiple stakeholders, reinforcing its role as a catalyst for broader systemic change.

Drawing on comparative case studies from across Europe, we aim to demonstrate the conditions under which design creates tangible value—not only in business performance but in social cohesion, identity-making, and sustainable growth. This study adopts a qualitative, exploratory research approach grounded in comparative case study analysis, aiming to investigate how different territorial and organizational contexts integrate design into their development strategies. The selected cases—Designregio Kortrijk, PROUD, Luxinnovation, and Cité du Design—were chosen based on their alignment with key themes in sustainable innovation, public–private collaboration, and cultural transformation through design. This selection allows for cross-case comparison while acknowledging contextual specificity, highlighting patterns and divergences in how design is mobilized as a strategic resource into value creation, into the operationalization of design at multiple scales—from cities to national agencies.

2 Towards Organizational Growth

Design is increasingly recognized as a source of competitive advantage for both companies and territories, identified by the European Community as a crucial bridge between creativity and innovation [5–7]. This relationship calls for a redefined perspective on innovation—one that emphasizes proximity, collaboration, and sensitivity to historical, cultural, political, and institutional specificities. Zurlo et al. [8] demonstrates that organizations with developed "design capabilities" outperform their peers not only in innovation outcomes but also in their adaptability to systemic challenges.

When design is associated with originality, creativity, and innovation, interpreting it as a set of management tools draws attention to the ways creativity can be systematically guided through planning and control. In this context, design becomes intertwined with strategic models and policies, shaping decision-making processes. In today's landscape of rapid technological advancement and economic globalization, promoting design has become essential for building identity and adding value across a range of assets, from individual products to regional and national brands. Within a highly competitive global market, design emerges as a critical means of differentiation, both in processes and outcomes.

Public policies and programs that foster design, involving collaboration between public and private actors, have laid the groundwork for diverse creative processes and outputs. From an organizational standpoint, design acts as a foundational element for development and a pathway to sustained growth. This relationship has been notably emphasized by the World Economic Forum, through initiatives such as the *"Global Agenda Council on Design and Innovation 2012–2014,"* and reports like *"Design for Innovation and Growth—a promising competitive concept in the future?"* [9]. These documents highlight key milestones in understanding the role of design in economic development, including a database of geographically and chronologically diverse examples. Although not exhaustive, the studies consistently show that "enterprises that use design have a better performance over time than enterprises that do not use design" [9].

3 The Innovation *Metaphor*

Within an economic and societal context that emphasizes results and outcomes, innovation has increasingly become associated with design culture. Through its practices, design offers a creative vision that supports strategic implementation. Innovation, in this extended perspective, is understood as a process of generating new opportunities and solutions, closely connected to creativity and idea generation, and applied to areas where practical value is recognized.

In the current organizational landscape, innovation has emerged as a central theme in strategic discussions. However, the role of innovation as a driving force for new solutions evolves alongside historical developments. As highlighted, "'Innovation' is one of those words that suddenly seem to be on everybody's lips, (…) how to design policies that stimulate innovation has become a hot topic at various levels of government" [10].

Innovation is promoted as a crucial factor for organizational development and sustainability, particularly when approached through systemic design that addresses complex societal challenges [11, 12]. The growing emphasis on innovation is visible across corporate, academic, and individual spheres. Innovation is defined as "the multi-stage process whereby organizations transform ideas into new/improved products, services or processes, in order to advance, compete and differentiate themselves successfully in their marketplace" [13]. This understanding promotes a multidisciplinary approach to innovation, framing it as a strategic process that connects knowledge, strategy, and partnerships. These components represent the foundation for interpreting innovation as a strategic imperative for organizational growth. In this context, design serves as a dimension for assessing innovation capacity, described by Laestadius et al. [14] as design intensity, where "a broad design concept including parts of what hitherto has been included in the design of R&D" is considered.

Design thus assumes a central role in shaping organizational innovation, summarized under the broader concept of "design innovation." In the current era of rapid technological development and economic globalization, design contributes actively to building identities and adding value to local industries' products. In a highly competitive global market, it functions as a differentiation tool, influencing both processes and outcomes.

Mutlu and Er [15] discuss this relationship by framing design as "the core function of innovation, and innovation as the main driving force in the economy." They note that "'design innovations' comprise (a) the incremental novelties in the design of an existing product or service, or (b) radically new products or services obtained by design effort with no or minimal technical novelty" [15]. Nevertheless, they also emphasize the challenges in achieving a concise definition, reinforcing the importance of practical case studies to validate the convergence of design and innovation.

Design is increasingly embedded within organizational structures, becoming a key factor in shaping knowledge, strategy, and partnerships. Junginger and Christensen [16] reposition the discourse around design and innovation by focusing on organizational culture, asserting that "design as a culture of making is seen to have a role in the making of culture" [16].

This integrated perspective is reflected in initiatives like the *"Design for Europe"* Action Plan (2014–2017), which links design to innovation by emphasizing user-centered approaches that create solutions with tangible value, notably through trends such as service design [17].

These perspectives support the idea of design occupying an institutional role, serving as a metaphor for innovation. However, the term "design" has also been subject to trivialization within public discourse, often conflated with other popular concepts such as creativity and innovation. Successful development models tend to be replicated across various industries, markets, and regions—a phenomenon that can be interpreted through Ritzer's notion of "McDonaldization," applied here to design strategy.

4 On *Creativity* and Transforming Creative Practices

Creativity is intrinsic to human existence and understood as the ability to pro- duce something that is both novel and useful. It relates to imagination and originality, with focus on problem solving or fulfilling a specific need.

It comes to vision from various combinations of individual and cultural capabilities, i.e., from the relationship between individual and collective. Runco and Jaeger [18] synthetize a standard definition of creativity from an extensive literature review, so the concept is thus considered as an asset or resource that must be managed and strategically developed, then valid to be incorporated in design strategy. It relates to a symbolic and emotional side to communication, breaking with conventional expectations and developing new visions, ideas and results.

Parkurst [19] acknowledges a confusion and lack of consensus in the definition of creativity. The author states it as a construct and proposes the following definition: "The ability or quality displayed when solving hitherto unsolved problems, when developing novel solutions to problems others have solved differently, or when developing original and novel (at least to the originator) products." [19]

In this approach, creativity also becomes a process of innovation, combined with multiple dimensions: ideas, skills, technology, and management, with which it defines itself as argument. It becomes the expression of territorial, social, individual and symbolic values as cause and effect for the desired shifts. We assume that creativity is a criterion of interest examining thinking skills, motives, and dispositional characteristics, among other variables.

5 A Sense of Place into Value Creation

Creativity is intrinsic to human existence and is commonly understood as the capacity to produce outcomes that are both novel and useful. It encompasses imagination and originality, often directed toward problem-solving or addressing specific needs. Creativity emerges from the interplay between individual capabilities and cultural influences—that is, from the dynamic relationship between the individual and the collective.

Runco and Jaeger [18], through an extensive literature review, synthesize a standard definition of creativity, framing it as a resource that can and should be strategically managed and developed. In this sense, creativity is validly incorporated into design strategy, contributing symbolic and emotional dimensions to communication, challenging conventional expectations, and generating new visions, ideas, and outcomes.

However, as Parkhurst [19] notes, there is considerable confusion and lack of consensus regarding the definition of creativity. Defining it as a construct, Parkhurst proposes: "The ability or quality displayed when solving hitherto unsolved problems, when developing novel solutions to problems others have solved differently, or when developing original and novel (at least to the originator) products" [19]. In this framework, creativity is positioned as a process that underpins innovation, integrating multiple dimensions such as ideas, skills, technology, and management. It expresses territorial, social, individual, and symbolic values, serving both as a cause and as an effect of desired transformations. Creativity, therefore, can be examined as a criterion of interest, encompassing thinking skills, motivations, dispositional characteristics, and other influential variables.

5.1 Focusing on Design and Place

Cities, regions, and countries are increasingly positioning themselves as territorial arenas of competitiveness and creativity, a trend that has accelerated with the rise of the knowledge society and, subsequently, the creativity paradigm.

Within this framework, design-led policies are expected to foster interventions across various spheres of economic, social, and cultural life, particularly within urban contexts. Only through a strategic articulation of these practices and processes can a genuine "design culture" emerge. By adopting governance strategies that integrate design, numerous cities have been recognized as "design cities" under the broader concept of the "creative city." As Landry [20] explains, this notion is grounded in "the idea that culture as values, insight, a way of life and form of creative expression, represents the soil from within which creativity emerges and grows and therefore provides momentum for development" [20].

The creative city model, while widely adopted, has also faced criticism for over-reliance on symbolic capital and underdeveloped inclusivity strategies [2]. This has prompted calls for more grounded, participatory design approaches [21].

Traditionally, the role of design in this dynamic has been associated with physical and iconic interventions—what could be described as a hardware perspective. Urban regeneration strategies often extend beyond architecture and infrastructure to include the proliferation of "gourmet culture," frequently linked to design, and the staging of ephemeral events such as festivals and exhibitions. These initiatives foster diverse social environments, marked by new consumption behaviors, lifestyles, and symbolic capital, contributing to the creation of renewed identities for places. However, such transformations are sometimes controversial, generating new interactions among residents, visitors, and entrepreneurs. As emphasized, "Considering creativity as key driver of both business and local development means in some way to put people at the core" [22].

Although typically associated with metropolitan centers, this strategic approach is increasingly pursued by small and medium-sized cities and regions aiming to restructure and diversify their economies while enhancing urban quality through design-led strategies. In this context, design emerges as a discipline that intersects economic, sociocultural, and environmental dimensions, playing a crucial role both conceptually and organizationally. Recognizing that intangible factors such as creativity and innovation are essential, design demonstrates that technology and quality alone are insufficient for achieving differentiation and sustainable growth. Contemporary economies rely heavily on creative and innovative competencies, underpinned by the belief that culture contributes significantly to value creation. Governance actors are increasingly acknowledging the role of creativity in economic strategy planning, not only by strengthening a city's social fabric and cultural diversity but also by nurturing a shared community identity.

As a multidisciplinary activity embedded in numerous processes within organizations, the knowledge inherent to the design process spans the entire value chain of a city or territory, reinforcing its critical role in contemporary development strategies.

6 Case Study Analysis

Therefore, this discussion is advanced through the benchmarking of a selection of case studies that explore the dynamic interplay between Sustainable Innovation, Entrepreneurship, and Technology, aligned with the interests and motivations of both organizations and citizens.

The analysis draws upon a set of dimensions that offer comprehensive and varied interpretations of the relationship between design, creativity, and organizational growth, serving also as a bridge to the theoretical frameworks that follow. In line with trends highlighted in the *New European Bauhaus*, initiatives like these demonstrate how multi-stakeholder design strategies can address urban regeneration, cultural identity, and innovation simultaneously. Following [23, 12] frameworks, the analysis focuses on five comparative dimensions: strategic integration of design, governance model, stakeholder engagement, tangible outcomes, and transferability. Table 1 presents a revised synthesis to support structured comparison.

6.1 Designregio Kortrijk

Designregio Kortrijk is a strategic partnership aimed at positioning the Kortrijk region (Belgium) as a hub that leverages design for its development. Its mission is to foster connections with other cities and regions across Europe, promoting design as a key strategic component.

Kortrijk and the surrounding Flanders region have undergone a remarkable process of renewal and creative transformation. Historically linked to the flax and textile industries, the region's economic base remains rooted in textiles, with the industry continuing to play a significant role. Since the early twentieth century, the region has increasingly shifted towards more advanced industrial activities, particularly in interior design and production sectors requiring technological and design solutions.

In 2005, to further accelerate this transformation, regional stakeholders in design established "Design Region Kortrijk." Since its inception, Designregio Kortrijk has operated as a joint initiative between the City of Kortrijk, the Interieur Foundation, the

Table 1 Case study analysis

Case study	Designregio Kortrijk (Belgium)	PROUD (EU)	Luxinnovation (Luxembourg)	Cité du Design (France)
Focus	Industrial revitalization	Innovation ecosystems	National design strategy	Urban and cultural regeneration
Governance	Multi-stakeholder: city, university, chamber of commerce	EU-funded collaboration of design centers and municipalities	National innovation agency	Public (municipality + national government)
Key strategies	Co-design, citizen participation, community branding	Co-creation, knowledge exchange, service design	National Action Plan for Design, SME support services	Biennales, research labs, public engagement
Impacts	Regional identity renewal, SME innovation	PTransnational knowledge transfer, designer support	Strategic design integration in companies, design literacy	UNESCO Design City designation, international design visibility
Challenges	Sustaining engagement, scaling results	Institutional fragmentation	Private sector buy-in	Gentrification risks, balancing artistic and social goals

Industrial University Howest, the Intercommunale Leiedal, and Voka—West Flanders' Chamber of Commerce. The organization coordinates the efforts of public and private entities, ensuring that design is deeply integrated into the region's industrial activities, educational infrastructure, and public sector initiatives. Through this approach, it reinforces the role of design in the region's development. As highlighted, "The forging of effective relationships between business and institutional stakeholders in innovation is an important mission for the City of Kortrijk in order to come to joint actions in the region" [24].

Participation plays a central role in this initiative, inviting creative professionals and organizations to expand their networks and engage in collaborative efforts. As part of this strategy, "To further sensitize the public to design, also highlighted the benefits for innovation and design of widening consumer input and contributions at the early stages of product development" [24]. One of the key outputs of this initiative is the Designregio community website, where individuals can create free profiles to share their work, projects, or calls for creative collaboration, reaching a broad audience.

6.2 Proud

PROUD—People, Researchers, Organizations Using Design for Innovation and Co-creation—was an initiative involving European organizations from various sectors, all united by their focus on design and innovation within their processes. The project was supported under the INTERREG IVB NWE program, an EU initiative aimed at promoting the economic, environmental, social, and territorial development of North-Western Europe. This program funded activities based on cooperation between partners from eight countries: Belgium, France, Germany, Ireland, Luxembourg, the Netherlands, Switzerland, and the United Kingdom.

The initiative was developed by eight partners from prominent design regions and cities: Eindhoven (Municipality of Eindhoven), Kortrijk (Designregio Kortrijk), Paris (APCI), Essen (DZNRW), Helsinki (Culminatum Innovation), Lancaster (ImaginationLancaster), and Luxembourg (Luxinnovation). Whether city councils, academic research institutions, design or innovation centers, or associations, all partners shared the belief that design is essential in addressing today's societal, ecological, and economic challenges.

PROUD aimed to bring together designers, public and private organizations, and user communities, bridging diverse needs to highlight the added value of design. The extensive experience of the partners led to the conclusion that new design approaches should be disseminated through hands-on learning. This approach also sought to support design studios, particularly small-scale organizations struggling to keep pace with material innovations, legal and economic regulations, and other developments impacting their profession.

This European project demonstrated that design and designers can make a significant contribution to "innovating, improving, and consolidating people's lives, businesses, and public services across Europe" [25]. It serves as a case study on how design can connect public authorities, SMEs, organizations, and citizens within a specific context, reflecting the unique characteristics of local organizational structures.

6.3 Luxinnovation

Luxinnovation is the National Agency for Innovation and Research in Luxembourg, founded in 1984 and established as an Economic Interest Grouping in 1998. Its primary objective is to consolidate national initiatives aimed at promoting innovation and research.

Luxembourg's unique political and geographical position enables it to serve as an attractive and dynamic hub for research and innovation. Thanks to its cultural tradition and recognition of the significant value of design, Luxinnovation positions design as part of a core triad that includes innovation, management, and creativity, while considering economic, technological, aesthetic, and societal factors. When design is strategically integrated into a company, it plays a vital role in generating innovative ideas and transforming them into marketable and viable products and services that meet user expectations.

To foster a design-driven culture among organizations and citizens, Luxinnovation introduced the National Action Plan for Design in 2009. This initiative offers a variety of services to clients wishing to incorporate design into their innovation activities. From a governmental perspective, the agency is tasked with driving design-related activities through a broad range of strategies, such as enhancing product functionality or creating new market.

6.4 Cité Du Design

Cité du Design in Saint-Étienne, France, inaugurated in January 2010, serves as a hub for research, education, communication, media, and design-related services. It caters to a wide range of audiences, blending urban and social activities with industrial and artistic expertise. This initiative is supported by a collaboration between the Municipality of Saint-Étienne, Saint-Étienne Metropole, the Rhône-Alpes Region, and the French State (Ministry of Culture). It stands as a public investment aimed at establishing a platform for higher education, research, economic development, and the promotion of art and design, accessible to diverse audiences.

The city of Saint-Étienne has a rich history tied to industrial modernity, driven by a synergy of art and industry. Through initiatives like the creation of the Cité du Design and the International Design Biennial, Saint-Étienne has embraced design as a catalyst for urban transformation and a key driver of economic, social, and organizational development. The city earned the designation of UNESCO City of Design in its inaugural year, a significant recognition of the importance and impact of its mission.

Cité du Design is a territorial project focused on education, research, and design experimentation, with its reach extending through the Saint-Étienne International Design Biennial. By emphasizing design as a key element of innovation in response to social transformations, it has developed research and experimental methods that have solidified its position on the international stage, making it a compelling case study for design's role in urban and societal development.

7 Closing Remarks

Design's relation with growth is rooted in the recognition of cultural knowledge, extending beyond the realm of experts and academia. This perspective presents significant challenges, particularly in terms of mediation, which serves as a cultural boundary. The observed interventions facilitate mobilization processes by creating platforms for dialogue and interaction. This study reaffirms design's role as a transformative and integrative force in the development of organizations and territories. Far from being limited to aesthetics or form, design—when understood as a cultural, strategic, and participatory practice · emerges as a mediator of value creation.

Design can play a pivotal role as a catalyst for organizational growth, especially when viewed through the lens of cultural knowledge validation [26]. In this broader framework, design transcends traditional boundaries—often confined by academic expertise and professional knowledge—and incorporates a deeper understanding grounded in cultural contexts and social dynamics. This shift compels organizations to engage in mediation processes that not only recognize cultural diversity but also foster connections across varying viewpoints. This paper argues that such connections do more than simply support growth; they actively shape it by promoting interaction and mobilization around shared goals and narratives, and the analyzed case studies demonstrate how design can foster economic resilience, reinforce community identity, and stimulate inclusive innovation, aligning with current calls for systemic approaches to sustainability and creativity [11, 12].

As design is positioned as an integrative and strategic function, its potential to serve as a unifying platform for dialogue and cultural expression becomes more apparent. This role facilitates the emergence of a creative economy that draws on cultural authenticity, collective experiences, and narrative-building. In doing so, design contributes to a dynamic organizational identity that can adapt to shifting cultural and market demands. Moreover, policymakers and practitioners should be encouraged to embrace design as both a method for inclusive governance and a driver of sustainable development [21, 23].

In this context, design's mediating capacity can be further understood through the lens of the helix models of regional innovation—particularly the *Triple Helix* (university–industry–government), *Quadruple Helix* (including civil society and media), and *Quintuple Helix* (adding the environment). Within these multi-stakeholder ecosystems, design functions as an integrative enabler, fostering collaboration across institutional and disciplinary boundaries. The case studies illustrate how design-led initiatives actively mediate between innovation actors by facilitating user engagement, shaping shared visions, and aligning local values with strategic development goals. As such, design becomes not only a cultural translator but also a system orchestrator, capable of enhancing knowledge co-creation, supporting democratic innovation processes, and grounding economic growth in local contexts and environmental responsibility. This positioning reinforces design's contribution to the symbolic helix of innovation, where material and immaterial resources intersect to create inclusive, adaptive, and sustainable development pathways [11, 12].

Acknowledgements This work was supported by FCT—Fundação para a Ciência e Tecnologia, I.P. by project reference UID/04057: Research Institute for Design, Media and Culture.

References

1. Oakley K, Banks M (2023) Cultural industries and the creative economy, 2nd edn. Routledge
2. Chapain C, Stryjakiewicz T, Kuebart A (2022) Creative cities revisited: critical perspectives on urban creativity. Eur Plann Stud 30(4)
3. Deforge Y (1995) Avatars of design: design before design. In: Margolin V, Buchanan R (eds) The idea of design. The MIT Press
4. Buchanan R (1995) Wicked problems in design thinking. In: Margolin V, Buchanan R (eds) The idea of design. The MIT Press
5. European Union: Design as a driver of user-centred innovation. Commission of the European Communities (2009)
6. European Union: Implementing an Action Plan for Design-Driven Innovation. Commission of the European Communities (2013)
7. European Union: New European Bauhaus—Beautiful, Sustainable, Together. Commission of the European Communities (2021)
8. Zurlo F, Cautela C, Rasca L (2023) Design capabilities for innovation and transformation. Springer (2023)
9. Kolmodin A, Pelli A (2005) Design for innovation and growth—a promising competitive concept in the future? ITPS, Swedish Institute For Growth Policy Studies
10. Fagerberg J, Verspagen B (2009) Innovation studies—the emerging structure of a new scientific field. In: Research policy, vol 38. Elsevier
11. Norman DA, Stappers PJ (2022) DesignX: addressing grand challenges with design. She Ji: J Design Econ Innovat 8(1)
12. Manzini E (2022) Designing in a complex world: social innovation, sustainability, and transformative systems. MIT Press
13. Baregheh A, Rowley J, Sambrook S (2009) Towards a multidisciplinary definition of innovation. In: Management decision, vol 47, no 8. Emerald Group
14. Laestadius S, Pedersen TE (2005) Towards a new understanding of innovativeness—and of innovation based indicators. In: Proccedings of low-tech as misnomer: the role of non-research-intensive industries in the knowledge economy. Brussels
15. Mutlu B, Er A (20203) Design innovation: historical and theoretical perspectives on product innovation by design. In: Proceedings of the 5th European academy of design conference. Barcelona
16. Junginger S, Christensen PR (2013) Design and innovation: organizational culture as making. In: Proceedings of 10th European academy of design conference—crafting the future
17. Dervojeda K et al (2014) Design for innovation. Service design as a means to advance business models—case study 13, Business Innovation Observatory, European Union
18. Runco MA, Jaeger GJ (2012) The standard definition of creativity. Creativity Res J 24(1)
19. Parkurst HB (1999) Confusion, lack of consensus, and the definition of creativity as a construct. J Creative Behav 33(1)
20. Landry C (2000) The creative city: a toolkit for urban innovators. Earthscan
21. Jégou F, Bonneau M (2022) Co-designing transition pathways in cities. Design Stud 79
22. Rivas M (2011) From creative industries to the creative place: refreshing the local development agenda in small and medium-sized towns. In: REDIGE, vol 2, No 2
23. Bason C (2023) Leading public design: how managers engage with design to transform public governance. Policy Press
24. De Clerck: Interview with Stefaan De Clerck, Former mayor of Kortrijk and President of Designregio Kortrijk. In: Innovation Pie, Year 2, Issue 4 (2011)
25. PROUD Europe: PROUD Positioning Paper 2.0. When co-design works: real change through collaboration
26. Brandão Pereira J, Quelhas V (2021) Organizational design towards the configuration of a media communication model. In: Markopoulos E, Goonetilleke RS, Ho AG, Luximon Y (eds) Advances in creativity, innovation, entrepreneurship and communication of design, lecture notes in networks and systems, vol 276. Springer Nature, pp 452–458

The Future of Social Design: Methodologies and Pedagogical Innovations: Building Community Capacity and Empathy in Design Practice and Education

Helena Grácio and Cátia Rijo

Abstract This paper examines how evolving methodologies and pedagogical innovations in design education can effectively address contemporary socio-economic challenges. Specifically, the study proposes a methodological framework that integrates social design principles into polytechnic higher education curricula to prepare designers who are both technically skilled and socially conscious. The research advocates for a shift from traditional problem-solving-oriented design methods to those centered on human experience and community engagement. The proposed framework emphasizes three core methodologies: co-creation, participatory design, and social innovation. These approaches are particularly suited to polytechnic education contexts, where practical application and industry readiness are paramount. The study demonstrates how educators can serve as facilitators who promote collaboration and student autonomy while preparing graduates for an evolving job market that increasingly values social impact alongside technical competence. Through theoretical analysis and pedagogical reflection, this research contributes a structured approach to integrating empathy, community capacity building, and holistic thinking into design education. The framework specifically addresses the unique characteristics of polytechnic institutions, where the balance between theoretical knowledge and practical application creates optimal conditions for implementing socially conscious design methodologies.

Keywords Social design · Collaborative methodologies · Pedagogical innovations · Community engagement · Empathy in design · Design education · Polytechnic education

H. Grácio (✉) · C. Rijo
Lisbon School of Education, Polytechnic University of Lisbon, Lisbon, Portugal
e-mail: hgracio@eselx.ipl.pt

C. Rijo
e-mail: crijo@eselx.ipl.pt

D. Raposo et al. (eds.), *Strategic Design for Sustainable Futures*, Springer Series in Design and Innovation 65, https://doi.org/10.1007/978-3-032-33282-0_9

1 Introduction

This paper provides a holistic, human-centered perspective on design methodology, developed within the context of design teaching practice in polytechnic higher education. Considering that the methodology used in design practice largely reflects the teaching of the discipline, we explore here the resources and processes that constitute the methodological approach employed. The aim is to offer an accessible and in-depth understanding of the involved processes, paving the way for education aimed at forming more empathetic designers who are aware of their social impact.

The purpose of detailing the different levels of methodological approach is twofold: first, to acknowledge the conceptual foundations that underpin an actionable model for design education in higher polytechnic education; second, to adapt this model, inspired by authors such as [16, 18], to the demands of a training program that prioritizes humanization and social impact. Thus, we propose an adaptation of the classical stages of design practice, adjusting them for an educational approach that balances technique with empathy and social responsibility.

Since the mid-twentieth century, design has been a discipline oriented towards problem-solving and creating solutions to contextual challenges. However, as design practice has evolved, so too have the methodologies that support this practice, moving beyond a purely rational and technical approach. In a post-war context, characterized by the need for economic reconstruction and accelerated growth, systems emerged to rationalize and systematize the design process, such as the proposals by [2, 4], which aimed to decompose problems into smaller parts to facilitate resolutions. However, this fragmented and technical view was challenged by more contemporary approaches that place the human being at the center of the process.

Today, an effective design methodology must be flexible and responsive to social demands, encompassing not only technical solutions but also emotional, cultural, and ethical dimensions. This vision is particularly relevant for design education, where it is necessary to prepare future professionals to address the complexity and interconnectedness of contemporary problems. This implies teaching not only technical tools but also socio-emotional skills, such as empathy, which Goleman [12] highlights as an essential competency for developing leaders and professionals capable of making a positive impact on society.

The design methodology proposed in this study, therefore, goes beyond the mere decomposition of problems and application of rational techniques. It seeks to integrate the development of human competencies, such as the ability to work collaboratively, understand users' needs and contexts, and act ethically. By adopting a more humanistic perspective, we recognize that the designer should be, above all, a mediator between technical knowledge and people's needs.

The proposed methodology is based on a flexible sequence of stages, which can be adapted according to the specifics of each project. This model reflects the growing complexity of issues faced by contemporary designers, who must deal with constantly evolving social, cultural, and technological variables. Unlike the purely technical approaches of the first generation of design theorists, such as Alexander and Broadbent, we advocate that active participation of various stakeholders in the process is fundamental. This point is supported by authors such as [17], who advocates for a social design approach, where all are active participants in the innovation process.

In the practice of design education in polytechnic context, we propose an approach that balances methodical structuring, such as that of [18], with openness to creativity and collaboration. Thus, the stages of the design process are seen not as rigid blocks but as flexible moments of reflection and action, where the focus is on deeply understanding the problem to be solved. This flexibility allows the designer to maintain a global and systemic view while keeping in mind the specific details required by each project.

In higher polytechnic education, where the focus is on practical and applied development, this methodology should incorporate elements of empathy and human-centered design from the early stages of the teaching process. We believe that by promoting a pedagogy that prioritizes engagement with the real needs of people, we prepare future designers to act ethically and responsibly. This model is supported by a methodological approach that balances technique with empathy, innovation with sustainability, and creation with critical reflection.

In summary, the methodological proposal presented here aims to form designers who are more aware of their social role, enabling them to act ethically and holistically. Based on a rich theoretical tradition that includes authors such as [16–18], we propose a teaching model that transcends technique and aligns with contemporary social and cultural demands. By integrating empathy, collaboration, and systemic thinking into design education, we believe it is possible to train professionals capable of contributing significantly to building a fairer and more sustainable future.

2 Methodology

As referred, the present article provides a comprehensive overview and reflection of the integration of collaborative and methodological approaches in social design and design education, highlighting the importance of a holistic and adaptive framework for both practice and teaching.

2.1 Foundations for a Model

By addressing the problems and sub-problems identified in the project, the designer can advance to more targeted research of analogous references. From this analysis, the initial guidelines for the proper development of the design methodology emerge. By breaking down the initial problem into different parts or sub-problems, the designer can address each one independently, systematizing various solutions for each stage. This way, it is possible to integrate the solutions found, culminating in a comprehensive and definitive solution to the proposed problem.

Once the problems are properly identified, the designer enters the analysis phase and seeks solutions. Some of these solutions may already exist or be found in related areas, and it is at this moment that the designer should filter out irrelevant information or solutions that prove to be inadequate. Careful selection of the obtained data allows for a solid path to be outlined for the development of the project, particularly in practical and logistical terms.

Following the analysis phase, the experimentation stage begins. In this phase, with the prior selection of data, materials, and technologies, as well as the systematization of sub-problems, the designer tests new approaches to known issues, deconstructing prejudices and preconceptions. Experimentation allows for the emergence of new models and solutions, encouraging innovation in project development.

Munari [18] argues that experimentation, coupled with careful observation and analysis, significantly reduces the chances of error in the design process. Thus, the hypotheses initially raised are validated with a lower risk of failure. At this point, simplification and careful selection of solutions become crucial, eliminating what is unnecessary and optimizing the time and resources invested in the project. This efficient process reduces costs and accelerates execution, while also eliminating confusion that could arise during development.

Understanding the various parts involved in the project, it is essential to ensure coherence in constructing the solution. Only with a harmonious approach is it possible to ensure that all parts of the project work in harmony, leading to the final stage of Munari 's methodology, inspired by Descartes' method: the verification of data and critical analysis of revisions that emerge throughout the process (Fig. 1).

In this stage, the designer compiles a detailed "recipe," containing all the elements and procedures necessary for the project's realization. It is the designer's responsibility to communicate this information through sketches or what Munari refers to as "constructive drawing." this stage prepares the ground for the production and execution of the final solution.

While describing his methodology, Munari develops a flowchart that evolves as he describes each phase of the process. This approach reaffirms the objectivity and rationalism applied to design (Fig. 1).

Lobach [16] complements this view by proposing that the design process is both creative and objective in problem-solving. In his methodology, Löbach divides the design process into four distinct phases (Fig. 2), highlighting that, although separate, these phases are interdependent and do not follow a linear order. The process is dynamic, allowing for advances and setbacks according to the nature of the project.

Löbach's first phase is problem analysis, where the designer gathers information and defines objectives. The second phase is generating viable alternatives, selecting methods, and creating ideas to address the problem. The third phase involves evaluating alternatives, carefully analyzing options to begin the selection process. Lobach [16], suggests that this evaluation should focus on two fundamental criteria, according to the product or project objectives. Finally, the fourth phase is executing the solution found, where, after gathering all data, the designer implements the idealized solution.

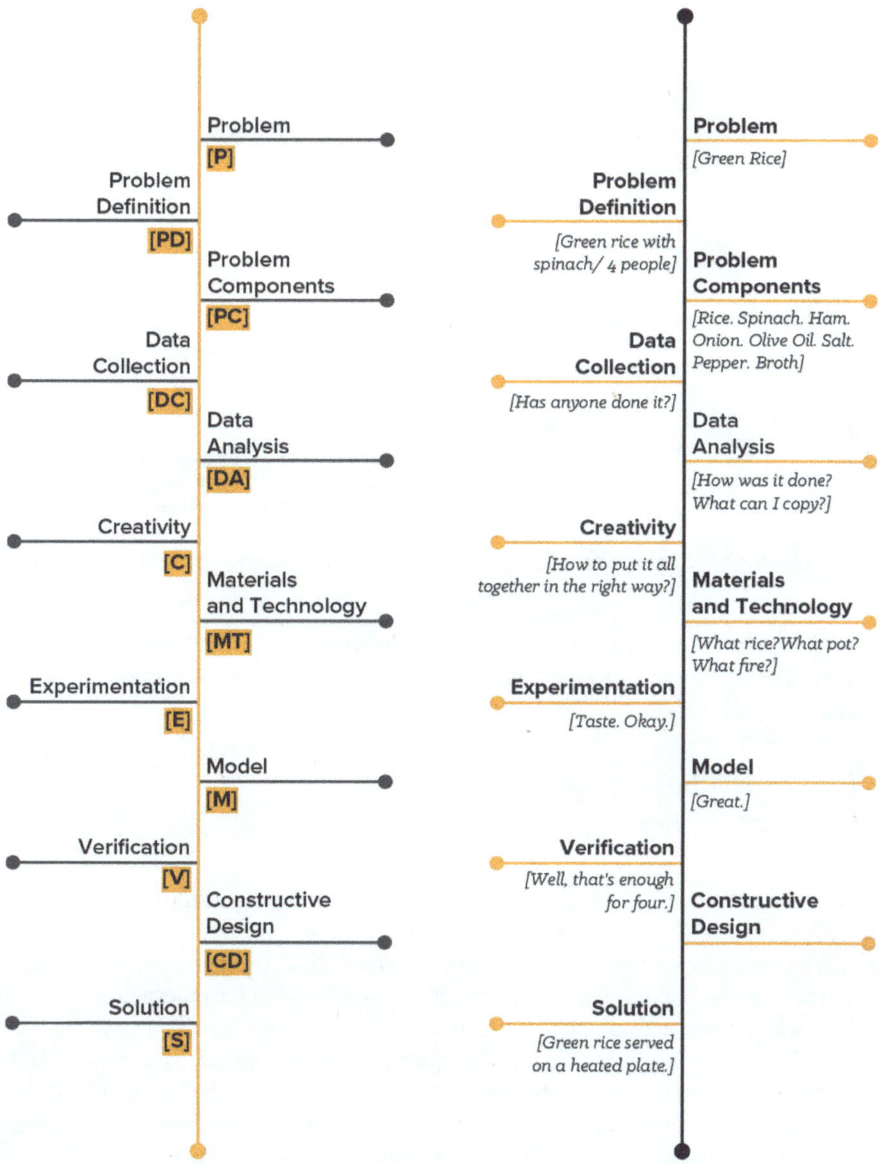

Fig. 1 Bruno Munari's project methodology proposal, adapted from Munari [18, p. 32]

Lobach [16], emphasizes that these stages are neither rigid nor sequential. The definition of influencing factors and goals occurs simultaneously with the definition of objectives and the search for creative solutions.

Several other authors have also addressed design methodologies throughout the twentieth century, proposing different approaches. However, in this research, the

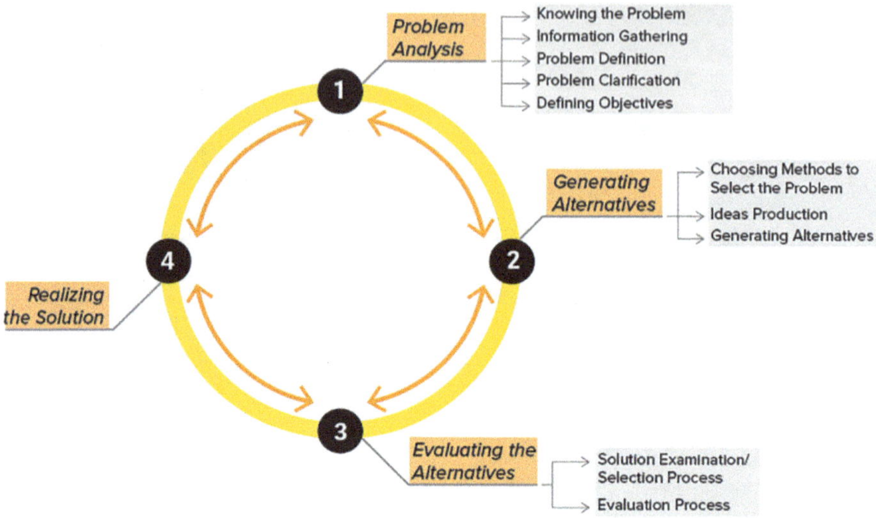

Fig. 2 Löbach's project methodology proposal, adapted from Lobach [16, p. 142]

methodologies of [4, 16, 18] are considered fundamental for the formulation of the adopted methodological strategy, guiding the development of the proposed action model.

2.2 *Social Design*

In the post-industrial era, as highlighted by [5], there has been a shift from an industry-focused society to an "information society" and a "knowledge economy," where the service sector and experience are the main drivers of the economy. This shift brought about the need to redefine the role of design, especially with the emergence of social design as a relevant and innovative practice.

Social design, as an emerging concept in the post-industrial era, focuses on collaborative processes and creating interactions that promote social change. Unlike industrial design, which primarily focused on creating physical products, social design involves conceiving solutions that engage human interactions, with the primary focus on improving social well-being and creating human and social value, as pointed out by [1].

The world design organization (WDO) defines design as a strategic problem-solving process that generates innovation and success, and that improves quality of life through innovative products, systems, and services. From this perspective, the designer must analyze and synthesize relevant data, working transdisciplinarily to meet the needs of producers and consumers.

According to Armstrong et al. [3], social design does not aim for objectivity and is not strictly solution oriented. Instead, it prioritizes human relationships and values, placing people and their experiences at the center of the design process. This implies that the designer must act as a mediator, facilitating collaboration between various stakeholders and ensuring that the outcome responds to the needs and aspirations of the community.

Designers in the social sphere must, therefore, address complex social and cultural issues, working with different actors to create solutions that are sustainable and effective in improving social conditions. This involves a shift from a product-focused approach to one that considers the social impact and collaborative potential of design. The evolution from traditional design approaches to social design reflects a broader understanding of the role of design in contemporary society, recognizing that the impact of design extends beyond the creation of physical objects to include the enhancement of human experiences and societal well-being.

The proposed methodology aims to reflect this understanding, integrating humanistic values and social responsibility into the design process. This approach intends to train designers capable of addressing complex social issues, ensuring that their work has a positive and transformative impact on society.

2.3 *Collaborative Processes and Action Models*

Success in social design relies on effective collaboration between social designers, existing social entities, and other stakeholders. Social design practice should consider community processes and the concept of community building through trust [23]. Designers should focus on building community capacities and aligning leadership through tools such as action research and organizational transformation.

Action research and organizational transformation provide essential platforms for social design, enabling improved interaction within the designed scenarios. Manzini [17] suggests that creating action plans to integrate participatory design and social innovation is crucial (Fig. 3). Designers in the post-industrial society must adopt methodological approaches such as co-creation, participatory design, and social innovation in their design processes.

Binder and Michelis [7] highlights the importance of working on a small scale and using various applications of disciplinary practice. Co-creation allows for working in workshop contexts with a wide range of tools, while participatory design involves fieldwork with consumers. Social innovation, on the other hand, employs a laboratory perspective linked to real scenarios, pushing designers out of their comfort zones.

The design market has evolved, resulting in a decrease in traditional employment opportunities and an increase in the pursuit of new markets and complex social challenges. This scenario has contributed to the development of new methodologies and a more holistic and inclusive approach in design [10]. Therefore, designers need to identify and understand collaborative processes and act as facilitators and mediators of social innovation, as design practice and the training of future designers

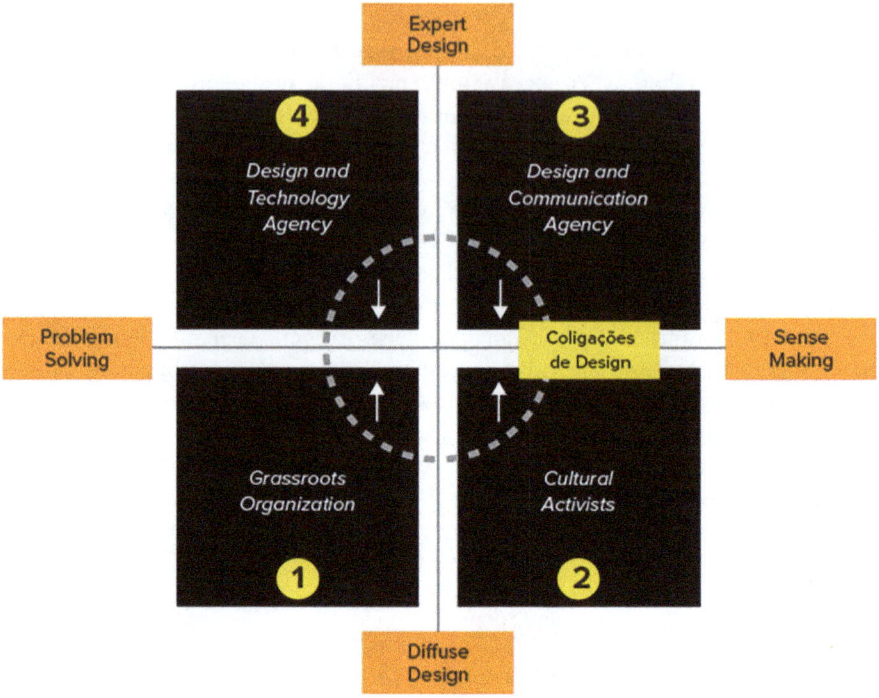

Fig. 3 Innovation Model proposed by Manzini, adapted from Manzini [17, p. 50]

should incorporate methodologies such as co-creation and participatory design to achieve social innovation.

In summary, social design is crucial in the current socioeconomic context, particularly in the knowledge economy and information society.

2.4 Methodological Approaches in Design Education

The twenty-first century has brought a greater diversity of student profiles in higher education, presenting challenges for educators in creating content that meets students' needs and prepares them for a competitive global market [13]. It is essential for students not only to master their field of expertise but also to gain confidence in applying knowledge in various situations. Educators should develop strategies that integrate practical and theoretical teaching, preparing students for the knowledge society [21].

In the landscape of learning institutions, like polytechnics, the teaching of design is at a moment where theory and hands on application must come together to nurture imaginative thinking geared towards creativity, innovation and problem solving.

While theoretical understanding serves as the basis, for any field of study the teaching of design requires a hands-on approach in polytechnic environments where the goal is to equip students to directly meet the demands of various industries. Emphasizing practice-based learning, can offer students a grasp of design principles that equips them to tackle real life obstacles and create innovative and practical solutions.

In design field—as a discipline—is primarily focused on applications which involve creating prototypes and testing them through hands on experiences to refine the process further at institutions where students are encouraged to engage directly with various materials and tools, along with digital technologies, to gain a firsthand understanding of bringing their designs to fruition. Engaging with the hands-on aspects of design can help students understand how theory translates into real world applications—a skill, for those pursuing careers in design fields.

Therefore, design education methodologies should promote a culture of memory through impactful pedagogical experiences. Considering empathy, a crucial tool in project development, as it enables the creation of user-centered solutions [9]. Assuming that we expect students, as future designers, to be able to engage with, and adapt to the contexts in which they work, in order to achieve the best design solution to the problems they are presented with, we believe that providing them with a vision in which they are able to create engagement with the context for which they work, it is necessary to equip them with a vision of empathy and a deep understanding of the community's culture, values and traditions [20].

Design practice should actively involve the community and focus more on the development process rather than specific products, enhancing the ability to design for the benefit of the community [14]. Design education should prepare students for real-world practice, including the transition to real contexts and teamwork [15, 22].

In higher education, the lecturer should act as a facilitator, encouraging student interaction and autonomy. A good facilitator should define the process clearly, allowing different agents to contribute their perspectives and skills [19]. Developing empathy and understanding the context is essential for translating cultural perceptions into contemporary solutions.

3 Results

In design education, outside the classroom setting as in polytechnic institutions fosters connections with local communities to engage students in designing solutions for actual societal needs on a practical level that prioritizes empathy driven design principles through meaningful interactions with users and consumers to address socio cultural and environmental aspects along with ethical considerations, in their work. This interaction contributes to forming a view of how design influences society and highlights the significance of collaborating with end users to guarantee that the solutions are genuinely suitable, for their intended purposes.

Ultimately, in schools the practical method to teach design is more than a teaching strategy; it is essential for shaping aspiring designers to face the real challenges of their profession. Through hands on learning experiences transferring knowledge through practice sessions engaging with communities and prioritizing the needs of end users design programs in institutions can provide students, with the necessary resources, abilities and perspectives to thrive in a demanding and ever-changing industry. As the world evolves with design influencing sectors and communities in ways the focus, on hands on learning will continue to be a crucial element, in preparing the upcoming designers of tomorrow.

Considering the reflection and analysis carried out so far, we propose the following model, rooted in social design and design education, highlighting the following key points:

1. **Effective Collaboration**: Social design should foster collaboration between designers, social entities, and other stakeholders.
2. **Community Capacity Building**: Focus on building community capacities and leadership through action research and organizational transformation.
3. **Methodological Integration**: Essential methodologies include co-creation, participatory design, and social innovation.
4. **Adaptation to Market Changes**: The evolving design market requires new methodologies and a holistic approach.
5. **Empathy in Education**: Design education should integrate empathy and real-world practice.
6. **Facilitator Role**: Educators should act as facilitators, enhancing student autonomy and understanding of community contexts.

Thus, this reflection points out that a thorough study of collaborative and methodological processes in social design reveals the imperative need for an integrated and adaptive approach that addresses contemporary challenges. The proposed action model highlights the importance of effective collaboration among designers, social entities, and other stakeholders, emphasizing the building of community capacities and aligning leadership through tools such as action research and organizational transformation. This model not only proposes a framework for social design practice but also illuminates the path towards a more holistic and inclusive practice.

4 Conclusions

The adoption of methodologies such as co-creation, participatory design, and social innovation emerges as essential for effectiveness in social design. These methodological approaches enable designers to work in a more collaborative and empathetic manner, adjusting to the dynamics and needs of the communities they engage with. The ability to adapt and innovate in response to changes in the design market is crucial, as it provides designers with new tools and methodologies that enrich practice and broaden professional scope.

This reflection underscores the necessity for an integrated and adaptive approach to social design, emphasizing effective collaboration, community capacity building, and methodological innovation. The model proposed offers a framework for social design practice and education, advocating for the incorporation of participatory techniques and empathy to achieve impactful social outcomes.

In the realm of design education, incorporating practical experience and empathy into the curriculum is crucial for preparing students for real-world challenges. Educators can play a pivotal role in facilitating student learning, guiding them to apply their knowledge in diverse and meaningful contexts. As the design field continues to evolve, embracing these practices will be essential for fostering positive change and addressing the complex social issues of the future.

References

1. Albino C (2014) Os Sentidos do Lugar: Valorização da identidade do território pelo design (Tese Doutoramento em Design). Universidade de Aveiro—Departamento de Comunicação e Arte
2. Alexander C (1964) Notes on the synthesis of form. Harvard University Press, Cambridge
3. Armstrong L, Bailey J, Jocelyn J, Kimbell L (2014) Social design futures. https://mappingsocialdesign.org/2014/10/09/social-design-futures-report. Accessed 6 Feb 2017
4. Asimow M (1962) Introduction to design. Prentice-Hall, Englewood Cliffs
5. Bachman L (2016) Sustainable design and postindustrial society: our ethical and aesthetic crossroads. ARCC J 13(2):30–38
6. Bell D (1974) The coming of post-industrial society. Harper Colophon Books, New York
7. Binder T, De Michelis G et al (2011) Design things. MIT Press, Cambridge
8. Broadbent G (1979) The development of design methods. Design Methods Theor 13:41–45
9. Brown T (2009) Change by design: how design thinking creates new alternatives for business and society. HarperCollins
10. Chen D, Cheng L, Lu-Lin C, Hummels C, Koskinen I (2015) Social design: an introduction. Int J Des 10(1):1–5
11. Descartes R (2006) Discourse on method (I. Maclean, Trans.). Oxford University Press (Obra original publicada em 1637)
12. Goleman D, McKee A, Waitz A (2017) Empathy (HBR Emotional Intelligence Series). Harvard Business Review Press, Boston
13. Gosper M, Ifenthaler D (2014) Curriculum design for the twenty-first century. In: Curriculum models for the 21st century using learning technologies in higher education. Springer, New York, pp 1–14
14. Julier G (2010) La cultura del diseño. Editorial Gustavo Gili, Barcelona
15. Kvan T (2000) Collaborative design: what is it? Autom Constr 9(4):400–415. https://www.sciencedirect.com/science/article/pii/S0926580599000254. Accessed 20 Mar 2019
16. Löbach B (2001) Design industrial: Bases para a configuração dos produtos industriais. Editora Blücher, São Paulo
17. Manzini E (2015) Design, when everybody designs. An introduction to design for social innovation. MIT Press, Cambridge
18. Munari B (2010) Das coisas nascem coisas. Edições 70, Lisboa [1st ed. 1981]
19. Retegi A, Sauvage B, Predan B, Tomás E, Schosswohl G, Kaltenbrunner M, Draganovská D (eds) (2019) The co-create handbook for creative professionals. Creative Region, Linz
20. Thomas J, McDonagh D (2013) Shared language: towards more effective communication. Australas Med J 6(1):46–54. https://doi.org/10.4066/AMJ.2013.1596
21. UNESCO (1998) World declaration on higher education for the twenty-first century: vision and action. UNESCO, Paris. https://unesdoc.unesco.org/ark:/48223/pf000014195

22. van Leeuwen J, van Gassel F, den Otter A (2004) Teaching collaborative design. In: Proceedings of the international workshop on construction information technology in education. Istanbul, Turkey, pp 1–13
23. Yee J, White H (2016) The Goldilocks conundrum: the 'just right' conditions for design to achieve impact in public and third sector projects. Int J Des 10(1):7–19

Design as a Means of Valuing Heritage and Sustainable Regional Development

J. Neves and R. J. Silva

Abstract Nature tourism and the promotion of heritage are fundamental elements in sustainable regional development, especially in territories with low population density and rural areas. This article evaluates the contribution of two re-search projects, the project "Ordo Christi—Artistic Heritage of the Order of Christ between the Zêzere and Tagus Rivers (fifteenth and sixteenth century)" and the project "Tourist-cultural and heritage signage in the Tagus International Natural Park", which aimed to explore the valorization of heritage and sustainable regional development, in the context of peripheral regions. For the results obtained in both projects, the application of multiple research methods and techniques was decisive, based on a mixed methodology, interventionist and non-interventionist, with a qualitative and quantitative basis. Based on the results obtained in the two research projects presented, it was possible to validate the role of design and the means developed for valuing the region's material and natural heritage and, consequently, for regional development.

Keywords Design for territorial development · Heritage valorization · Regional development · Order of christ · PNTI (Tejo Internacional Natural Park)

1 Introduction

Higher education institutions today have a determining role in the promotion of territories, especially in peripheral or low-density regions. This mission goes beyond traditional teaching and research activities, seeking to generate a direct and positive

J. Neves (✉)
Research Centre for Architecture, Urbanism and Design, CIAUD, Lisbon School of Architecture, University of Lisbon, and IPCB, Polytechnic University of Castelo Branco, Castelo Branco, Portugal
e-mail: joaoneves@ipcb.pt

R. J. Silva
School of Arts and Humanities, ARTIS - Institute of Art History, University of Lisbon, and IPCB, Polytechnic University of Castelo Branco, Castelo Branco, Portugal

D. Raposo et al. (eds.), *Strategic Design for Sustainable Futures*, Springer Series in Design and Innovation 65, https://doi.org/10.1007/978-3-032-33282-0_10

impact on society, contributing to the economic, social, and cultural development of communities.

As a center of knowledge and innovation, the academy seeks to directly benefit the regions by transferring the results of research, development, and technologies to local companies and organizations, contributing to innovation and increased competitiveness, benefiting regional economies.

Design, as a discipline oriented toward problem-solving and the creation of human centered and context-sensitive solutions, plays a fundamental role in developing innovative responses for territories. Through design, it is possible to develop visual, communicational, and functional strategies that contribute to the valorization of local assets, promote cultural identity, and reinforce a sense of belonging. Design acts as a catalyst for social and cultural innovation, serving as a mediator between technical knowledge and empirical wisdom, between tradition and contemporaneity, and between economic development and environmental sustainability.

In line with its institutional mission of qualification, knowledge production, and dissemination, the Polytechnic University of Castelo Branco, through its Higher School of Applied Arts, has played an important role in cultural, artistic, and technological training, but also in terms of research, development, and innovation. There are countless applied research and experimental development projects carried out in the territory, valuing culture, heritage, local identity, sustainability, and society.

In the interior territories of Portugal, marked by depopulation, aging populations, and economic fragility, applied research and experimental development are fundamental tools for generating value. These territories require solutions tailored to their specificities, which can only emerge from close collaboration, community involvement, and interdisciplinary experimentation. The production of knowledge in real contexts, oriented toward action, has enabled the development of participatory methodologies, signage systems, territorial identity projects, and cultural mediation products that help revitalize the local social and economic fabric.

The projects "Ordo Christi—Artistic Heritage of the Order of Christ between the Zêzere and Tagus Rivers" and "Tourist-cultural and heritage signage in the Tagus International Natural Park" are representative of this approach. Both operate at the intersection of research, heritage, and design, proposing solutions that respect the historical, cultural, and natural values of the territories where they are implemented. The first project has contributed to giving visibility to significant artistic heritage, reinforcing intermunicipal cohesion and promoting new cultural routes. The second has provided a protected area with an effective orientation system that improves the visitor experience and fosters environmental education. In both cases, the impacts go beyond material results, producing applied knowledge, strengthening collaborative networks, and mobilizing local resources for more cohesive, inclusive, and sustainable development.

This article results from the evaluation of the contribution of two research projects developed in the central region of Portugal, specifically in the Beira Interior subregion, which aimed to explore heritage enhancement and sustainable regional development within the context of peripheral regions. The "Ordo Christi—Artistic Heritage of the Order of Christ between the Zêzere and Tagus Rivers (fifteenth and sixteenth centuries)" project focused on the survey, promotion, and valorization of tangible heritage, while the "Tourist-cultural and heritage signage in the Tagus International Natural Park" project aimed to create an information and guidance system to improve visitation conditions and promote natural heritage.

Based on the results obtained from these research projects, this study sought to understand the role of design and the means developed in the valorization of the region's material and natural heritage, and consequently, in regional development. The investigation focused on how the studied local heritage and the implemented solutions could help the region and its communities achieve balanced and responsible economic and social development, preserving resources and the environment for future generations—particularly in peripheral or low-density regions, where the challenges and opportunities are distinct.

1.1 The Ordo Christi Project

Following the dissolution of the Order of the Temple decreed by Pope Clement V in 1312, King D. Dinis of Portugal institutionalized, in 1319, through the papal bull Ad ea ex quibus issued by Pope John XXII, the Militia of Our Lord Jesus Christ. This new military order incorporated the former knights of the Templar Order [1, 99–132], as well as all their privileges and assets—such as castles, villages, and lands—whether under direct Templar rule or merely sponsored by the order. The Templar presence deeply shaped and marked the national territory, particularly in border or frontier zones. These cultural traces remain clearly visible today, both in material and immaterial forms, and much of their preservation is owed to the legacy assumed by the knights of the Order of Christ [2, 82].

The new order, Ordo Militiae Jesu Christi, consolidated its political standing and patrimony throughout the fourteenth century, particularly after 1420, when King João I obtained from Pope Martin V the authorization to appoint his son, Prince Henry the Navigator, as administrator of the Order of Christ (1420–1460). This correlation of powers gained increasing geopolitical significance, notably during the reigns of King D. Manuel I and his son King D. João III, when the patrimony of the Order was significantly consolidated. That vast patrimony was manifested in multiple artistic expressions sponsored or commissioned by the Order: immovable heritage such as castles, churches, manor houses, pillories, and housing; and movable heritage resulting from numerous commissions across various commanderies, reflecting its importance in the territory, especially between the end of the Middle Ages and the beginning of the Modern period. The presence of the Order of Christ in the territory is clearly marked by the jurisdictional organization of the commanderies. As noted by Isabel Sousa e Silva, a "Comenda" is defined as "a delimited territorial space within which the knight friars exercised lordly authority in the name of the Master and according to the power dele-gated to them, benefiting from its goods and revenues. Organized as lordships and de-fined as agrarian management units, they combined income from the land with a broad set of territorial and jurisdictional rights."

By 1505, within the geographic scope of this project, sixteen comendas had been identified: Alcains, Alpedrinha, Bemposta, Castelo Branco, Castelo Novo, Escalos de Cima–Escalos de Baixo, Caféde and Aldeia da Mata, Idanha-a-Nova, Idanha-a-Velha, Lardosa, Lousa, Penha Garcia, Proença-a-Velha and São Miguel de Acha, Rosmanin-hal, Salvaterra, Segura, and Vila Velha de Ródão.

In all these places, the Order asserted its authority as landowners, which led to the accumulation of various types of property, as documented in the Tombos das Comendas (inventories) and the Visitações (inspections) of the Order of Christ. This wealth inevitably left deep and lasting marks. Today, that reality is still visible in the region's to-ponymy, where numerous references to the Order of Christ are found throughout the former territories of its commanderies, such as Rua da Quinta da Ordem (Fundão),

Quinta da Ordem (Vila Velha de Ródão), Quinta Souto da Ordem (Castelo Novo), Caminho da Quinta da Ordem (Fundão), Terra da Ordem (Monsanto), among others.

In addition to these preserved place names and collective memories, the visual presence of the Order's jurisdiction and spiritual authority still endures. Across the territories once under its control, one can easily and clearly identify the heraldic symbols of the Order carved into pillories, town halls, religious buildings, and residences of the Comendadores.

Although the region's patrimonial dimension—inevitably linked to the legacy of the Templars—is increasingly acknowledged today, the Beira Interior still lacks a global and integrated vision of this historical-artistic heritage. Nevertheless, the growing body of historical and artistic studies published in recent years, the research projects conducted or underway, and the efforts of municipalities, dioceses, and museums within this geographical area to study, systematize, and disseminate their heritage are all important steps toward this broader understanding.

The project "Ordo Christi—Artistic Heritage of the Order of Christ between the Zêzere and Tagus Rivers (fifteenth and sixteenth century)" was proposed by the Polytechnic University of Castelo Branco, with participating entities being the Polytechnic Universities of Guarda and Tomar and the tourist promotion company Naturtejo. Also participating in the project were the municipalities of Covilhã, Fundão, Penamacor, Idanha-a-Nova, Castelo Branco, and Vila Velha de Ródão, the Dioceses of Guarda and Por-talegre/Castelo Branco, and Artis—Institute of Art History of the School of Arts and Humanities at the University of Lisbon. The project was financed by the Centro Regional Operational Program 2020 and evaluated by the Foundation for Science and Technology (FCT).

The project contributed to the study, safeguarding, valorization, and promotion of the regional and local heritage inherent to the Order of Christ in the central interior of Portugal. The research work aimed to promote knowledge and contribute to making historiographic information clearer and more accessible to society, seeking its applicability for tourism and educational purposes. To pursue the proposed objectives, strategic tools were found in communication and information design for disseminating the results and routes developed, expressed through multiple means of information, dissemination, and promotion.

In addition to studying the historical-artistic heritage of the Order between the Zêzere and Tagus rivers, the project was deeply concerned with communicating, in an open and perfectly accessible way to everyone, the artistic heritage carried out under the authority of the Order of Christ. In this way, the project sought to be a memory and identity tool for the region, which, through various communication elements as well as the creation of heritage routes, effectively contributed to raising awareness among society, territorial cohesion and dynamization of the region through its history.

Order of Christ is important for territorial identity and cohesion. This set of material goods that make up the historical-artistic heritage (where those of an intangible nature can be added) are the result of creative activities and an aesthetic ideal [3, 4]. Its own value must be considered of relevant interest for the permanence of memory, a testimony to the identity of the culture of a society and its respective regions [5]. The inexhaustible value of heritage, although often fragile and perishable, requires combined synergies for the creation of memories, preservation, safeguarding and transfer of knowledge to be a counterweight to the gal-loping phenomenon of globalization.

1.2 The PNTI Signage Project

In recognition of its ecological importance, natural and scenic values, and with the aim of promoting nature conservation in harmony with human well-being, the Tejo/Tajo International Transboundary Biosphere Reserve was designated by UNESCO in 2016. The Tejo/Tajo Biosphere Reserve includes the Parque Natural do Tejo Internacional (in Portugal) and the Parque Natural del Tajo Internacional (in Spain).

On the Spanish side, the Parque Natural del Tajo Internacional extends across 11 municipalities in the province of Cáceres, within the Autonomous Community of Ex-tremadura: Alcántara, Brozas, Carbajo, Cedillo, Herrera de Alcántara, Membrío, Salo-rino, Herreruela, Santiago de Alcántara, Valencia de Alcántara, and Zarza la Mayor.

The Parque Natural do Tejo Internacional (PNTI), officially established in 2000 by Regulatory Decree nr. 9/2000, is a protected area located in eastern Portugal, encompassing the municipalities of Castelo Branco, Idanha-a-Nova, and Vila Velha de Ródão, within the district of Castelo Branco (Fig. 1).

The border section of the Tagus River and the deeply embedded valleys of the Tagus and Erges rivers are characterized by their rugged and wild landscape, with steep escarpments, rocky gorges, and predominantly schistose geological formations. This natural diversity supports Mediterranean biotopes such as cork oak and holm oak woodlands (montado), cereal steppes, and riparian gallery forests. It is one of the most important habitats in Portugal for the conservation of endangered bird species, including the black stork (the symbol of the Park), the cinereous vulture, the Spanish imperial eagle, and the golden eagle, among other species protected by international conventions. The Park's value is further enhanced by the presence of unique plant species and endemic freshwater fish. This biological richness, combined with low

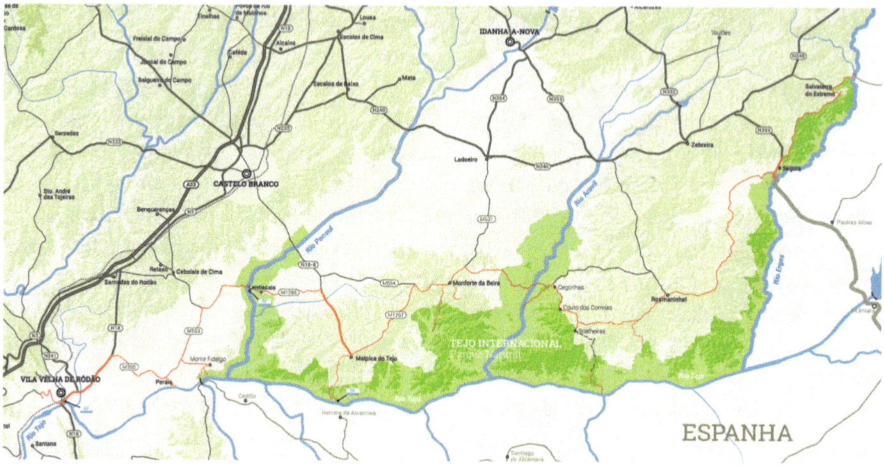

Fig. 1 Coverage map of the International Tagus Natural Park, including the municipalities of Idanha-a-Nova, Castelo Branco and Vila Velha de Ródão

population density and limited human pressure, makes the PNTI a priority territory for conservation strategies, environmental education, and sustainable valorization.

Like other natural Parks in Portugal, the PNTI has historically suffered from under-funding and a centralized, restrictive management model imposed by national authorities. In response, the Ministry of Environment and Energy Transition, through Order no. 4237/2018, initiated a process to implement a new intervention model for protected areas. Within this framework, the Pilot Project for the Collaborative Management of the Tejo International Natural Park was launched in 2018—an innovative initiative that embraced a new form of protected area management based on participation and collaboration, as defined in the National Strategy for Nature Conservation and Biodiversity.

The Pilot Project for Collaborative Management involved seven partners: the Institute for Nature Conservation and Forests (ICNF), the municipalities of Castelo Branco, Idanha-a-Nova, and Vila Velha de Ródão, the Polytechnic Institute of Castelo Branco, the Beira Baixa Business Association, and Quercus—National Association for Nature Conservation. This initiative marked a turning point in territorial intervention, establishing a culture of collaboration and participation in the Park's management (PNTI) while simultaneously fostering a new dynamic focused on community involvement and sustainable development.

Within the scope of the Pilot Project, several initiatives were developed, including the project "Tourist-cultural and heritage signage in the Tejo International Natural Park" (PNTI), coordinated by the Polytechnic Institute of Castelo Branco, with the mu-nicipalities of Castelo Branco, Idanha-a-Nova, and Vila Velha de Ródão as beneficiaries. The project was funded by the Environmental Fund, as stipulated in the Order of the Office of the Minister for Environment and Energy Transition no. 4538/ 2019.

This project was part of the 2018–2022 PNTI Enhancement Plan, which outlined 20 action measures across four strategic axes. The signage project was aligned with Axis 1—Communication, creation and promotion of identity, specifically Measure 3.1—To create and maintain a network of orientation equipment to support tourism and leisure that improves visitation conditions within the PNTI. The goal was to improve mobility and circulation within the Park by installing vertical signage along access roads to points of interest with visitation value, providing visitors with guided, safe, and environmentally respectful movement throughout the protected area.

The central objective of the "Tourist-cultural and heritage signage" project was to enhance and improve visitation conditions within the Park through the implementation of a visual identity system and a coherent guidance and information system consistent with institutional communication purposes. The project also included the study and development of a graphic language for the Park's visual identity, capable of being applied across other media and used within the guidance and information system. This encom-passed three key dimensions: information, interpretation, and orientation for tourists and visitors within the territory (Fig. 2).

From an informational perspective, the aim was to develop a set of tools to promote and improve visitation conditions, including the designation of information points and tourist offices in each municipality to support decision-making for tourists and visitors and to provide digital, printed, or inperson information about points of interest, routes, infrastructure, services, and products available in the territory.

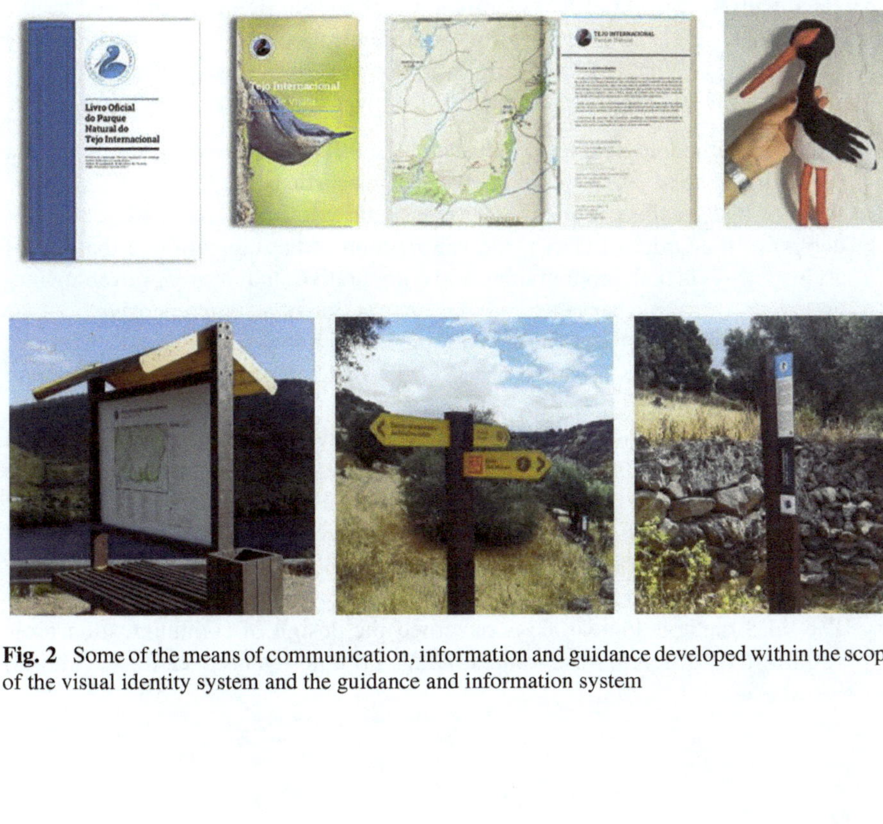

Fig. 2 Some of the means of communication, information and guidance developed within the scope of the visual identity system and the guidance and information system

From an interpretative standpoint, the project aimed to meet the challenge of designing and implementing a guidance system expressed through signage and interpretive supports. This included both analog and digital systems distributed throughout the Park—from its entry "gates" to areas of high natural and scenic value. Interpretative spaces were established in municipal centers, alongside information points in key locations, interpretative panels describing local fauna and flora, and other educational tools designed to promote environmental awareness and interpretation.

The initiative also aimed to meet the challenge of implementing a comprehensive guidance system articulated through specific signage and wayfinding elements. These included road signage and tourist-cultural signage, from panels along motorways to directional signs with geographical and ecological indicators. Additionally, specific signage was developed and applied to guide visitors through short walking routes and interpretive trails within the Park.

2 Methodology

Given the significance of the Order of Christ in the interior regions of Portugal and the extensive artistic legacy it produced, the project "Ordo Christi—Artistic Heritage of the Order of Christ between the Zêzere and Tagus Rivers (15 and 16th century)" set out to identify, study, systematize, and disseminate the historical and artistic heritage associated with the Militia of Jesus Christ throughout the Beira Interior land-scape. This vast geographic area extends from the northern boundary of the municipality of Covilhã to Vila Velha de Ródão, and from Penamacor to Castelo Branco.

The methodology of the Ordo Christi Project was structured across three levels:

(a) The first methodological level focused on identifying, surveying, studying, and producing content related to the historical and artistic heritage. This included the study of historical/artistic heritage associated with the "Comendas" (commanderies) of the Order of Christ; the valorization of heritage through the lens of art history—via both monographic and comparative studies of immovable, integrated, and movable heritage; and the construction of a comprehensive heritage corpus that would support the development of knowledge and an integrated vision of the Artistic Heritage of the Order of Christ between the Zêzere and Tagus rivers.
(b) The second level aimed to ensure articulation between scientific knowledge and its adaptation to the profiles of diverse stakeholders. It sought to contribute to the definition of tourist routes capable of generating value for the territory and fostering territorial cohesion; and to support the preservation of memory and the safeguarding of the historical and artistic legacy, which serves as an identity marker of local and regional history.
(c) The third methodological axis concerned the design of communication tools optimized for different stakeholders and their promotion in educational, cultural,

and touristic contexts. This involved the application of heritage communication strategies appropriate to contemporary society; contributions to civic and educational development through the formation of new audiences, especially via new technologies; the creation of promotional materials; the stimulation of sustainable tourism in the territory by overcoming municipal boundaries; and the fostering of economically relevant synergies among regional entities and stakeholders—reinforcing identity, knowledge, and territorial cohesion through heritage.

The first methodological step of the Ordo Christi project was to identify the immovable heritage, with a total of 84 properties being identified, divided into religious, military, and civil architecture. The research also focused on movable and integrated heritage that is associated with the Order of Christ. Subsequently, we proceeded to study and prepare content related to the aspect of historical-artistic heritage associated with the commendations of the Order of Christ. The mapping carried out (and underlying the mobile and integrated) immediately reveals the potential tourist value and the attractiveness of places and goods [6–9]. With these data, we set out to establish synergies between scientific knowledge and its adaptation to the profiles of the different stakeholders: Definition of tourist routes capable of generating value for the territory and promoting its cohesion; Collaboration in safeguarding heritage; contribution to the conservation of the memory of the historical-artistic heritage, which constitutes an identity mark of local/regional history.

Finally, means of communication were developed, optimized for the different actors and promoted in an educational, cultural, and tourist context. Construction of a corpus of images suitable for different audiences; Application of heritage communication strategies appropriate to society.

For the Tourist-Cultural and Heritage Signage Project at PNTI, a methodology divided into three phases was defined. The first phase focused on conducting an exhaustive territorial diagnosis. This stage involved a detailed analysis and survey of existing routes, natural, cultural, and heritage points of interest, as well as the available visitor support infrastructure. The inventory of existing signage allowed the identification of gaps and inconsistencies in the previous system, contributing to a more rigorous initial mapping. In addition, extensive fieldwork was carried out, including insitu data collection, photographic records, and assessment of accessibility and the conservation state of key sites. Based on this fieldwork, the data were updated and refined in line with the project's objectives, which led to a significant expansion in the number of signable points of interest and the optimization of the support infrastructure network for visitors.

The second phase of the project corresponded to the planning and technical design of the information and wayfinding system. In this phase, all routes, points of interest, and support infrastructures were validated in coordination with project partners, ensuring institutional and technical alignment with the involved municipalities, the Institute for Nature Conservation and Forests (ICNF), and other relevant entities. Based on this joint validation, the precise georeferencing of all signposting locations was under-taken using GIS (Geographic Information Systems) and digital mapping tools. Circulation plans were then defined, identifying the main access routes to the park, visitor zones, hiking trails, scenic viewpoints, and interfaces between natural areas and urban settlements. The design of circulation flows was guided by criteria such as safety, legibility, graphic coherence, and respect for the natural and scenic values of the protected area.

The third and final phase consisted of the design, development, and implementation of the signage and information system. This phase included the creation of graphic, interpretative, and informational content tailored to different user profiles, as well as the development of physical signage using durable and sustainable materials. The system included large-format panels installed along major roads, serving to identify and promote the Park, and close-range directional signage containing geographic and ecological information. Specific signage was also developed for short hiking trails and interpretive routes, including totems, landscape reading panels, plaques describing local flora and fauna, and educational supports located in strategic areas. This final phase also involved the production of a technical manual for the application of the signage system, ensuring its replicability, future maintenance, and institutional consistency in all communication within the Tejo International Natural Park.

3 Results

Tourism activity in Central Portugal has, over recent years, recorded growth rates above the national average, according to the most recent data released by the National Statistics Institute (INE).

In the specific case of the Beira Interior region, the figures confirm that, as in previous years, the upward trend in tourism in Central Portugal continues to attract visitors and generate positive impacts on the local economy. Between 2020 and 2024, according to the INE and Turismo do Centro, tourism in Beira Interior experienced a marked recovery across all municipalities covered by the projects analysed in this study, following the significant downturn caused by the pandemic.

Castelo Branco recorded an increase from approximately 45,000 overnight stays in 2020 to around 93,000 in 2023, representing a growth of over 100%, thus consolidating its position as one of the main regional hubs. The municipality of Covilhã followed a similar trend, almost doubling its number of visitors from around 99,000 to 184,000 over the same period. Fundão also showed significant growth, rising from just over 31,000 to 69,000 overnight stays, an increase of 121%. Likewise, Idanha-a-Nova saw a rise from 11,000 to 23,000, representing an increase of more than 110%. Although on a smaller scale, both Penamacor and Vila Velha de Ródão accompanied this trend, effectively doubling their figures: Penamacor from approximately 6,000 to 12,000, and Vila Velha de Ródão from 2,800 to 6,200 overnight stays.

Taken as a whole, the region demonstrates remarkable dynamism, with all municipalities significantly surpassing their 2020 figures. This confirms a strengthening of the region's touristic appeal, rooted in its natural resources, historical heritage, and cultural events. The diversity of the tourism offer—which includes the potential of the Historical Villages and Schist Villages, adventure tourism and local gastronomy, the region's eco-tourism, and natural landscapes [10]—continues to attract both domestic and international tourists, as evidenced by tourism reports and the indicators published by the INE.

Of the Ordo Christi Project resulted in the open and free availability of scientific and appropriate informative tools in graphic language; heritage study, to be made available on the internet and also on analogue media, notably on paper and tourist support systems; creation of Tourist Routes; contribution to civic and educational training, with the formation of new audiences mainly through new technologies; Creation of multi-platform website, maps, information and guidance system, info-graphics, videos and animated images; Contribution to building sustainable tourist solutions inherent to the territory, thus diluting municipal borders and fostering synergies with economic relevance for regional entities and stakeholders; Provision of content to participating or interested entities, especially tourist promoters and municipalities, so that they can be applied to the valorization, dissemination and tourism promotion of the region; Publication of scientific information in open access.

On the other hand, the Ordo Christi Project promoted a global and informed vision of the immovable and integrated heritage of the Order of Christ, with the hope that the results will be enhanced through a communication strategy open to the participation of municipalities, dioceses, the Naturtejo Geopark, hotels, restaurants, tourism promoters and other interested agents.

Fundamentally, one of the main results is the contribution of the Ordo Christi Project to the promotion of heritage tourism and local development, through routes properly structured itineraries in time and space, as described by [11] anchored in territorial-based strategies that, through the construction of networks and points of interest, enable the holistic promotion of local resources with high belonging and attraction value.

Additionally, the project consolidated a methodological approach that values applied research and multidisciplinary collaboration, bringing together fields such as Art History, Communication Design, Cultural Tourism, and Territorial Planning. The collaborative construction of the heritage corpus among researchers, municipal technicians, and local communities served not only as a tool for cultural valorization but also as a means to reinforce collective identity and a sense of belonging. Through the systematization and graphic visualization of information, the project made previously scattered or inaccessible knowledge available, thus democratizing access to heritage and fostering its appropriation by local populations.

It is also important to highlight the project's impact on enhancing the strategic positioning of the Beira Interior region within the national and Iberian contexts. By integrating the heritage of the Order of Christ into a logic of intermunicipal and transregional heritage networks, the project laid the groundwork for future dynamics of institutional cooperation, particularly within European programs for cultural valorization and sustainable development. The dissemination of results in academic and professional settings—through conferences, publications, and digital platforms—strengthened the recognition of the territory as a space of memory, innovation, and sustainable tourism promotion.

With the implementation of the Tourist-Cultural and Heritage Signage Project at the PNTI, it was possible to create and promote the maintenance of a network of equipment and infrastructures to support tourism and leisure that improved visiting conditions at the PNTI, notably the mobility of visitors due to the availability of adequate signage for access to points of interest. It was possible to publicize the conditions for visiting the Park, as well as its natural heritage.

The results of the Project made it possible to improve circulation levels in the PNTI, through the provision of an information and guidance system installed on the access roads to points of interest and with visitation value in the conservation unit. In this context, two complementary systems were applied, one relating to Tourist-cultural and Heritage Signage and the other relating to Signage Models for purposes of identification and information relating to nature and biodiversity conservation in the National Network of Protected Areas.

Based on the results obtained in the two research projects presented, it was possible to validate the role of design and the means developed for valuing the region's material and natural heritage and, consequently, for regional development. The questions investigated aimed to analyze how the local heritage studied and the solutions found can create conditions for the region and communities to achieve balanced and responsible economic and social development, preserving resources and the environment for future generations, especially in these peripheral or low population density regions, where the challenges and opportunities are distinct.

In addition to improving physical orientation throughout the territory, the project reinforced the visual identity of the PNTI through a unified and institutionally coherent graphic language. This visual identity, consistently applied across all physical and digital media, contributed to the symbolic valorization of the park and to its perception as a space of ecological and touristic significance. The consistency of the visual communication was key to establishing institutional credibility with visitors, tourism stake-holders, and local communities, fostering a sense of belonging and ownership of the territory.

Simultaneously, the project had a significant impact on environmental education and awareness-raising for nature conservation. The creation of interpretative materials—such as informational panels on local fauna and flora, landscape reading points, and ecological signage—facilitated the understanding of the natural values present and encouraged responsible visitor behavior. The implemented system thus became an open-air educational tool, mediating the relationship between the public and the park's sensitive ecosystem, promoting sustainable tourism practices, and strengthening the connection between conservation, knowledge, and the valorization of natural heritage.

4 Conclusions

Numerous challenges and obstacles were encountered throughout the development of these two research projects. Both initiatives are characterised by a particularly significant territorial scope, involving multiple municipalities and addressing a range of strategically relevant dimensions for their respective regions. These factors contributed to generating interest among various political stakeholders and fostered the creation of synergies between the different actors involved.

The agents operating on the ground proved essential to the successful development of the projects. Given the often-complex logistical requirements and the need for extended periods of fieldwork—including observation, mapping, and the collection of photographic, video, and descriptive data—the researchers relied on the support of several facilitators, particularly in gaining access to the various study sites. This support was crucial in achieving the objectives set out.

Although both projects have been completed and yielded valuable academic results, it cannot be said that their full implementation has been realised. In the specific case of the Ordo Christi project, despite the significant outcomes achieved at the research level, it remains unimplemented, even in light of the clear interest expressed by the municipalities involved. This is largely due to the current political priorities in place.

By contrast, the project aimed at promoting and enhancing the Tejo International Natural Park has been successfully implemented. It comprises a visual identity system, as well as a wayfinding and information strategy, supported by a range of promotional materials and resources designed to facilitate and enrich the experience of tourist visits. From the identification and study of artistic heritage, the Ordo Christi Project contributed to solving problems in a territory, in a process of desertification, at risk of underdevelopment and with low cultural and tourist attractiveness, in which investment in culture, innovation and territorial cohesion are fundamental. In view of these problems, the Ordo Christi project had a reach that goes beyond the contribution it can make to the history of art, seeking to be a tool to stimulate public and private investment [12].

The PNTI Project has the great virtue of encouraging expanded institutional mobilization for specific intervention in the territory, enhancing the combination of efforts, knowledge and capabilities in a dynamic of positive sum for nature and the region.

The projects contributed—through their approaches—to systematization but, above all, to the dissemination of heritage, based on communication design, tourism and economic strategies. Only this way is possible to strengthen the complex webs of memory that maintain the identity of communities and the preservation of their history. Sharing knowledge, through concerted and adequate communication, allows us to achieve capacity to enhance and generate regional wealth, combining, for example, with modern concepts, which are also not unrelated to the Ordo Christi and PNTI projects.

In this way, Heritage communication cannot be hermetic, but rather accessible and easy to understand, where society reviews its cultural and natural heritages and promotes its continuity over time in face of globalization of globalization. In this context, the historical-artistic study of the Order of Christ or the International Tagus Heritage allows us to reflect retrospectively on local and regional history in order to build a future based on memory and cultural and natural heritage, but above all to review ourselves in these projects identity marks and how it becomes visible, in so-called modern societies, the identity factor is a central and defining issue for territories, especially peripheral territories.

Based on the results obtained, it is considered that nature tourism and the promotion of heritage are fundamental elements in sustainable regional development, especially in territories with low population density and rural areas. These segments promote the appreciation of the region's natural and cultural resources, bringing economic and social benefits without compromising the environment and local culture.

Ordo Christi and the Tourist-cultural signage project for the PNTI proved to be fundamental for strengthening the region's tourism offering and for building a more attractive and better-structured territory. The heritage routes created, the signage systems implemented, and the interpretive and informational materials developed provided improved visitation conditions and promoted a richer and more informed experience of the territory. These actions contributed to the creation of meaningful tourism experiences with strong differentiation potential at the regional and national levels, fostering new economic dynamics, reinforcing intermunicipal collaboration, and enhancing tourism's role as a driver of sustainable development.

In the field of innovation and knowledge, both projects emerged as exemplary practices of applied research, oriented toward solving concrete problems and supported by interdisciplinary methodologies. By intersecting fields such as art history, communication design, geography, planning, and conservation sciences, they opened new possibilities for studying and valorizing heritage in real-world contexts. They also generated educational and scientific resources with practical utility and social impact, including manuals, publications, digital platforms, and pedagogical materials. These results rein-force the role of academia as an agent of innovation at the service of communities and validate the importance of collaboration between higher education institutions, municipalities, businesses, and civil society organizations.

Finally, it is important to highlight that these projects contributed to rethinking the role of culture and heritage as strategic development factors, especially in territories affected by remoteness, demographic aging, and economic fragility. They demonstrated that heritage valorization goes beyond material preservation and involves activating its meanings, engaging communities, and creating territorial value. Through the articulation of memory and future, science and creativity, these projects embodied a contemporary and integrated vision of heritage, affirming it as a platform for identity, social cohesion, and sustainable regeneration.

References

1. Lencart J (2016) As Ordenações inéditas da Ordem de Cristo de 1319 e 1323—estudo comparativo com as Ordenações de 1321 e 1326 [The unpublished statutes of the Order of Christ of 1319 and 1323—Comparative study with the statutes of 1321 and 1326]. População e Sociedade 26:99–132
2. Gomes SA (2011) A extinção da Ordem do Templo em Portugal [The extinction of the Order of the Temple in Portugal]. Revista de História da Sociedade e da Cultura 11:75–116. Imprensa da Universidade de Coimbra
3. Muñoz Viñas S (2003) Teoria contemporanea de la restauración. Editorial Sintesis
4. Riviere GH (1993) La Museología - Curso de museologia / Textos e Testimonios. Akal
5. Tugores F, Planas R (2006) Introducción al patrimonio cultural [Introduction to cul-tural heritage]. Ediciones Trea
6. Cominelli F, Greffe X (2019) L'économie politique du patrimoine culturel [The polit-ical economy of cultural heritage]. In Situ. Au regard des sciences sociales (1). https://doi.org/10.4000/insituarss.436
7. Greffe X (2003) La valorisation économique du patrimoine [The economic valorization of heritage]. La Documentation Française
8. Greffe X (2011) L'économie politique du patrimoine culturel: de la médaille au rhizome [The political economy of cultural heritage: From medal to rhizome]. In: ICOMOS 17th general assembly. Paris, France. http://openarchive.icomos.org/1307/1/IV-3-Article6_Greffe.pdf
9. Pérez MD, Fernández JM (2015) El patrimonio cultural, recurso estratégico para el enriquecimiento económico y social: ejemplos desde el Patrimonio Mundial en España [Cul-tural heritage, a strategic resource for economic and social enrichment: Examples from World Heritage in Spain]. In: Personas y comunidades: Actas del Segundo Congreso Inter-nacional de Buenas Prácticas en Patrimonio Mundial [People and communities: Proceedings of the Second International Congress on Best Practices in World Heritage]. Universidad Complutense de Madrid, pp 777–792
10. Quinta-Nova L, Ferreira D (2024) Analysis of the suitability for ecotourism in Beira Baixa region using a spatial decision support system based on a geographical information system. Reg Sci Policy Practice 16(1). https://doi.org/10.1111/rsp3.12750
11. Ramírez JH (2011) Los caminos del patrimonio: Rutas turísticas e itinerarios culturales [The paths of heritage: tourist routes and cultural itineraries]. Pasos. Revista de Turismo y Patrimonio Cultural 9(2):225–236. https://doi.org/10.25145/j.pasos.2011.09.019
12. Nypan T (2004) Cultural heritage monuments as value generators. Directorate of Cultural Heritage

Sustainable Innovation in Textile Design: Applying Artificial Intelligence to Enhance Local Heritage

Eugenia Chiara and Ana Fernandes

Abstract This study explores the convergence of artificial intelligence and the creative process of textile design within a fashion education project inspired by Italian, European and global local cultural heritage. An experimental approach was adopted, integrating research and design development phases, combining artificial intelligence algorithms with subsequent manual processing, to generate fabric samples. The study is based on a careful analysis of local cultural heritage (shapes, symbols, colours), reinterpreted in a contemporary and global way. The results show how artificial intelligence can expand creative possibilities while requiring a critical and conscious interaction from designers. The conclusions highlight the potential and key aspects of the combination of artificial intelligence and creative design and propose an operational model to enhance territorial and cultural heritage through an innovative and interdisciplinary visual language.

Keywords Sustainable innovation in design · Artificial intelligence in fashion design and textile pattern design · Design education within the triple helix · Design and cultural heritage

1 Introduction

This study considers the way artificial intelligence (AI) could be utilized in fashion and textile design practice as a culturally embedded, sustainable innovation tool. As an alternative to AI as a human imagination replacement, we suggest AI as an instigator—a one which can help in the reinterpretation of cultural heritage using new visual lexicons and digital discovery. Based on the Triple Helix model, focusing on

E. Chiara
Politecnico Di Milano, Milan, Italy

A. Fernandes (✉)
Polytechnic University of Castelo Branco, Castelo Branco, Portugal
e-mail: ana.fernandes@ipcb.pt

D. Raposo et al. (eds.), *Strategic Design for Sustainable Futures*, Springer Series in Design and Innovation 65, https://doi.org/10.1007/978-3-032-33282-0_11

collaboration between academia, industry, and territorial stakeholders, the research positions design education as an innovation platform embedded in local identity.

The emphasis is placed on designing an educational system that unites students to work alongside generative AI tools with a critical appraisal of the role such can play in the preservation and reterritorialization of cultural heritage. This demands not only technical investigation, but ethical reflection, cultural awareness, and compatibility between automation and human beauty. AI is consequently a means for enriching context-specific local material and symbolic processes in innovation-oriented and context-sensitive ways. The project contributes to broader arguments concerning sustainable innovation—not only environmentally, but also the capacity to maintain cultural diversity, community control, and ethical design processes within a globalized creative economy.

The research particularly gives attention to textile pattern design, a sector where the combination of generative AI and human craftsmanship manual refinement can create innovative new possibilities for visual narratives that are rooted in the local identity.

This research moves in a four-phase methodology. The first phase consists of literature examples that look thoroughly through the current usage of AI in fashion sector. These are drawn from recent academic contributions from Dhaenens, Esposito and Willie [1], Gupta [2], and others. These resources examine AI's impact on creativity, consumer behaviour, education and the transformation of cultural identity. The second phase is mainly an analysis of important software platforms and case studies with the target of finding out the emerging tools and their importance to heritage-driven projects and their evaluation of how these design initiatives are applicable in culturally important heritage projects.

The third phase focuses on the analysis on an experimental educational framework experimented during the" Italian Lifestyle and Fashion" course held at the Politecnico di Milano in the first semester 2024–2025 by the professors Arturo Dell'Acqua Bellavitis and Eugenia Chiara.

The course integrated a research phase and a projetual phase that integrated AI technologies into the textile design process, creating motifs inspired by different local heritage, in Italy and other countries, such as Sicilian dream, House interior in Apulia, Meltemi mood, Balcanic interior.

The main goal of the project was to develop textile motifs using AI, refining them using human abilities, while reflecting on both the technical process and the cultural narratives that they wish to convey. The resulting patterns and participant feedback are assessed qualitatively.

The final phase of the study discusses key insights, with attention to both the potential and the limitations of AI integration. It concludes by proposing pathways for ethical, creative, and culturally sensitive use of AI in fashion and textile design education and practice.

2 Problem

2.1 Research Topics

The overall issue this research addresses is the way artificial intelligence can be integrated into design approaches to valorise and reinterpret indigenous cultural heritage—i.e., that of fashion and textiles. As design education increasingly employs digital media, the question becomes how such tools can be integrated without weakening cultural specificity or creative ownership but rather amplifying the same.

Artificial intelligence promises unique possibilities of visual production and data-driven design. Yet in its application to heritage projects, AI also brings fundamental questions: How is content generated by AI culturally relevant? Where does human curation come in mediating the tension between automation and meaning? And how can learning in design be a testing ground for ethical, critical, and creative uses of AI in such an environment?

To respond to these inquiries, this study explores how design education can teach AI not only as a technological device but also as an ethical and conceptual one. The study explores:

- Where in the fashion and textile design process AI can contribute positively to creative development, particularly when inspired by heritage.
- How design educators can train designers to integrate AI well into heritage-inspired design processes.
- How much innovation spurred by AI can be channelled towards sustainable design principles—important not only in the environmental context, but also cultural continuation and geographical relevance.

Based on the Triple Helix approach, the book examines the ways in which academe, industry, and local cultural actors can collaborate to create technologically sophisticated yet culturally embedded systems of innovation. It places design education in the role of a sustainable hybrid innovation driver.

The project thus contributes to a broader agenda of design-led regional transformation, where sustainability is interpreted not only in environmental terms, but also as cultural continuity, social inclusion and responsible creative agency.

Parallelly because of didactic research the paper mentions some relevant case studies about platforms and projects using AI in the processes of graphic and textile design, fashion collection concept generation and visualization, trend research and retail.

2.2 AI, Fashion and Textile Design Process and Education

Starting point is a reflection about the role of designers in the fashion and textile project and how it can be redefined using ai.

"The role of a designer is much more complex than just generating images, it is about concept, references, subjectivity and also the technical skills" [3, p. 43].

"While AI applications have been present for quite a while in many fields, its contributions to the textile industry are fairly new" [4].

The State of Fashion Report by McKinsey and Company and the Business of Fashion [5] argues that "designers and merchandisers will be empowered to make faster decisions and steer brands towards a demand-driven model with 3D technology, virtual sampling and AI-supported planning".

AI can be a demand-driven and time reducing tool, but human guidance and intervention is essential to maintain a cultural and formal quality of the overall projects.

As design educators is important to approach the use of AI in Design Process in a critical and cultural way considering that:

"The first challenge is to educate about Artificial Intelligence, which translates—in the era of post-mediality, platformization and datafication—the classic objective of Media Education, that is, the promotion of critical thinking, the understanding of "what is behind" a cultural product, what messages and what interests" [6, p. 281].

2.3 Case Studies and Fields of Application of AI in the Design Process

Starting from these insights the course analysed some relevant cases of the use of Ai in four different phases of the fashion and textile process, mentioned in the list below:

- Graphic, visual and textile design artists (Refik Anadol, Mario Klingermann, Robbie Barrat, Liam Young, David Brudnkin, David Szauder, Scott Eaton, Arthur Machado, known as Tùúk'z, Joai, Obvius art, Zach Lieberman)
- The fashion AI creatives that often collaborated with well-known brands (@str4ngerthing, @paa.tiff, @masha_kozhdan, @aidesign.png, @irisvanherpen, @unmadeunmade, @subakuruonline, @taskin, @the_fabricant, @Crisstianaaff, @krista.kim, @opethestylist, @mangrovia_info)
- Trend Research Platforms as Trend Forecasting Platforms, Visual Trend Analytics, Design automation Platforms, social media Trends Analytics (Google trend, Euritech, TensorFlow and Paytorch, Apash Adope and Spark, Brandwatch, Nielsen, Kaggle, Stichfix, constructor.op, revolve, Style3d, 3dBook, Fendy beauty)
- Retail Innovative Platforms and Projects as Recommendation Engines, Visual Search and image recognition, AI Powered Inventory Management, Augmented Reality Fitting Rooms, Chatbots, Dynamic Pricing (Walmart Smart Cells, Sephora Virtual Artist, Nike's AI Chatbot, Amazon AI Power Recommendation System, Pinterest Lens, Hm Chatbot).

2.4 Platforms and Application of AI in the Design Process

Following the design experimentation carried out during the course is possible to affirm that AI can be used in the fashion and textile design process as powerful tool in some specific phases of design process, always guided by the human subjectivity and prior knowledge.

The students used and analysed many different AI platforms related to image and pattern generation such as Midjourney Adobe Firefley, Stable Diffusion, Dall-E, PatternedAI, LibLibAI, Creative Fabrica, Fotor, Wow Patterns, Xinva.ai, Phot.ai, Pattern Cafe, creating also tutorial for the pattern generation process (Figs. 1, 2, 3 and 4).

- During the rich design process these tools were used for many different functions such as user analysis (personas visualization);
- Concept definition (textual brief definition, conceptual image generation, mood-board visualization);
- Trend research and trendboard visualization;
- Graphic pattern definition;
- Textile/outfit contextualization and styling.

Fig. 1 AI generated personas project Sicilian Dream. Students: Martina Signoroni, Krush Patel, Ryley Liu

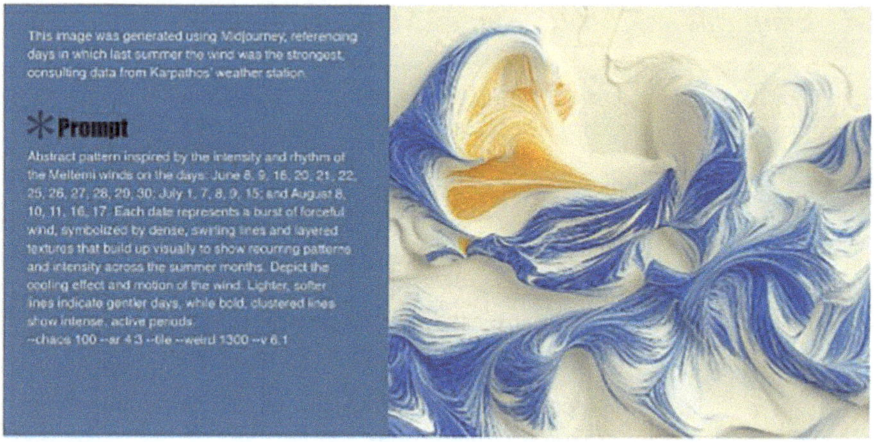

Fig. 2 Conceptual AI generated image project Meltemi Mood. Students: Yuhe Zang - Juliet Lagae—Marjolijn De Vil

Fig. 3 Ai generated pattern Student project Italian Architecture Inspiration. Students: Delehaye Louise, Monsef Alexandre, Paumelle Lou Wang Xin

Fig. 4 Ai generated textile contextualization project Gurajati Mood Students: Nandini Mehta, Anju Konikkara George, Huang Jingqi

3 Methodology

The method of this study is organized in a four-stage action-research cycle that integrates qualitative analysis, experimental pedagogy, and practice-based design-inquiry. It is assumed that heritage-based innovation in textile design must come through participative iteration between theory, technological investigation, and critical reflection [7].

Phase 1—Theoretical and contextual underpinning: Case study examination and literature review was undertaken to discover the state of play with respect to AI uses in fashion and textile design. Effort was focused on the relationship between AI, creativity, cultural identity, and learning. The theoretical underpinning of how AI was conceptualized as a creative force and an important problem of design pedagogy was established in this phase.

Phase 2—Mapping of AI tools and platforms: A comparative analysis of generative AI tools (e.g., Midjourney, Stable Diffusion, Patterned AI, etc.) was conducted in view of their possible application within design workflows drawn from local heritage. Four broad areas were mapped: pattern and visual creation, concept generation, trend research, and retail integration [8]. The aim was to recognize the way these tools could aid or obstruct culturally responsive design results.

Phase 3—Experimental education pathway: An experiment on a course module was conducted at Politecnico di Milano with more than 100 students engaged in AI-supported design projects. The approach was one of AI prompt engineering, cultural explorations, manual fine-tuning, and pattern application in context briefs. The phase experimented with the integration of AI in an actual pedagogical context, assessing both creative performance and educational impact.

Phase 4—Critique and critical synthesis: Qualitative examination of students' final work, design activity, and reflective statement was conducted. This phase engaged with evaluating the influence of AI tools on cultural representation, creative agency, and pedagogical engagement. Strategic recommendations were given through these findings regarding how to ascertain sustainable innovation in design education using AI, in the broader framework of the Triple Helix model of collaboration.

By uniting critical analysis with experimental use, this method allows for a situated appreciation of AI's use in fashion design, registering its capacity to facilitate culturally attuned, ethical, and pedagogically successful innovation.

4 Development

4.1 Analysis of the Experimental Educational Framework

This paper presents the modelling of an experimental educational pathway that merges design and research through the study of artificial intelligence (AI) applications in the field of fashion. With an emphasis on valuing cultural heritage, the course developed to explore the potential of AI in the conception and development of fashion and textile projects. Through the integration of creative experimentation with highly advanced digital tools, the pathway is both aimed to encourage critical thinking and technical innovation among students. The course's structure follows a multi-phase approach that is designed to guide the participants from the theoretical understanding to practical implementation.

The course" Italian Lifestyle and Fashion" took part at the Politecnico di Milano in the first semester 2024–2025 and was led by the professors Arturo Dell'Acqua Bellavitis and Eugenia Chiara. In the course they participated around 100 students from about 20 nationalities (italy, China, India, Azerbaijan, Australia, France, Lebanon, Egypt, Etiopia, Netherland…) that worked on 30 different briefs everyone related to a specific local context, such Sicilian Dream, Meltemi Mood, Danish Home, Alpine House for the development of a textile family of patterns generated with AI and inspired to the local material heritage.

The first phase focused on theoretical framework and case study analysis. The students were offered with an overall view of how AI is currently used in fashion. Graphic design, concept and collection development, trend analysis and retail strategies applications were included in it. The students were able to build a fundamental grasp of the technological and cultural implications of AI in fashion by studying real-world examples and currently emerging practices.

The second phase was solely dedicated to concept creation through AI. Activities included in this session was AI-assisted brainstorming using tools such as ChatGPT, benchmarking of existing products and services, company and user analysis (including the creation of personas), and the generation of synthetic images, concept boards, and moodboards all using generative AI. These exercises promoted students to explore AI not only as a support tool but also as a co-creator for the design process, which promoted innovative thinking and visualizing fashion concepts [9, 10].

The third phase, students moved to Ai-assisted pattern development, they started off with trend research and creation of colour palettes. Students designed original patterns and explored variations, experimenting with form, texture, and cultural references closely knit to local heritage using AI platforms.

The fourth phase mainly concentrates on the execution of pattern developments. This involves refining the AI-generated prompts that the students created along the experiment. They were also asked to create multiple variations of the pattern and create different applications for it in interior spaces. Iterative experimentation and contextual design thinking was given priority in this phase.

Finally, in the finalization phase of the project students were asked to rework manually on the generated graphics. Create a coherent family of patterns and apply the final output designs in interior spaces. This closed the loop between digital exploration and tangible application. The course was able to demonstrate how AI cannot be used to augment creative processes and for reinterpretation of cultural heritage in meaningful and innovative ways. This reinforces the importance of hybrid methodologies in contemporary design education.

In this context, the project illustrates a concrete example of academia–industry collaboration rooted in principles of sustainability and cultural innovation. Rather than a one-way transfer of technology, the initiative fostered a co-creative process in which artificial intelligence was critically employed to reinterpret local material and symbolic heritage.

This collaborative framework aligns with the Helix model of regional innovation, promoting synergies between academic research, production capabilities and territorial valorisation. The connection with local textile contexts enabled academic outputs to be grounded in real-world applications, encouraging design responses that honour cultural identity while opening pathways for responsible innovation within the fashion and textile sectors.

5 Results

The integration of AI in textile design based on heritage revealed greater creative potential alongside prevailing tensions between cultural specificity and algorithmic production. This study represents the outcome of a research based experimental course focused on the using of AI tools to create patterns for textiles inspired by local heritage of diverse countries. The course was part of an educational project which included both an analytical and a design phase. It engaged young fashion creatives in a practice-based research process. The project resulted in the creation of a mixed method incorporating experimental design, AI methods, and critical thinking—especially suited for fashion design using cultural heritage. It also insights into the collaborative potential of AI and human beings. It shows the limitations of AI in cultural interpretation.

The course was mainly organized into two phases: an initial analytical phase and subsequent design phase. During the analytical phase, students were asked to explore cultural heritage of selected regions through both conventional research methods such as historical texts, visual archives and interviews and AI-assisted tools which are capable of processing and synthesizing massive datasets of images and textual information. These tools were based on generative and machine learning models which are aided in identifying motifs, colours and symbolic elements with different cultural contexts.

During the design phase students were able to translate the insights that they gained into textile patterns using AI generative systems such as diffusion models to create visual prototypes. These AI tools were trained or fine-tuned with the help of culturally specific data prepared during the initial analysis phase. The co-creation process involved iterative feedback loops, which were evaluated, refined and reinterpreted AI outputs by the students through a human centric perspective. This interplay between algorithmic generation and perspective human curation became the very central part of the methodology.

The development of a new action-research educational methodology which is specific to the field of fashion design is one of the primary results of the project. By giving emphasis on this experimentation and iterative practice, this model is not just positioned towards students and designers but also for active researchers participants were able to critically reflect on how emerging technologies can inform, expand, and sometimes enhance, enlighten, and occasionally challenge their creative processes, through direct engagement with AI systems. The approach encourages the flexible roles between researcher and practitioner, enabling a dynamic and context-specific exploration of AI in fashion.

A second key outcome is regarding the concerns of the nature of insights provided by AI in fashion. AI might detect repeated visual patterns and come up with fresh mixtures beyond the reach of conventional intuition. Those works, although provocative, needed cultural screening and design refining by students to become relevant in their contexts. Ai also proved that it is particularly effective in identifying visual patterns. Moreover, AI helped to facilitate the abstraction of traditional elements into contemporary forms which supports a form of digital storytelling that respects the cultural roots along the same time enabling innovative ideas [11, 12]. This research confirmed the necessity of human interpretations at the key stages of the process. Human beings are essential to assess the cultural appropriateness of Ai-generated outputs, so that final patterns are ensured to remain respectful and meaningful within their specific heritage frameworks.

Finally, the research was able to identify the critical challenges and limitations in the applications of AI in the field of cultural heritage. AI tools seemed too often operate based on the data-driven logic that overlooks the subtle, emotive and non-verbal dimensions of cultural heritage. This risks in flattening aesthetics and the culturally diverse traditions are rendered into visually pleasing elements with shallow conceptual motifs. This highlights the ethical need of informed curation and continuous dialogues between designers, AI systems and cultural stakeholders. The most frequent limitation was the overrepresentation of Western-centric data sets, and these created aesthetic and cultural biases that tended to flatten non-Western symbolic depth. This helps to emphasize the need for culturally sensitive dataset curation [13, 14].

In conclusion, the experimental course was able to provide a value testing ground for understanding how AI can be integrated into fashion design processes centered on cultural heritage. The course was able to show that AI was able to demonstrate that it can act as a creative partner offering new perspectives for pattern generation.

The research was able to emphasize the irreplaceable role of human judgment, especially when dealing with sensitive and culturally embedded content. The research thus advocates the need for a hybrid, ethically-aware approach to AI in fashion—one that empowers designers to navigate the space between tradition and technology with critical sensitivity and creative confidence.

6 Conclusions

This article concludes the broader research effort initiative focused on the integration of artificial intelligence (AI) into fashion design projects that aim in valuing local heritage. Based on the outcomes of an experimental course which involved design institutions and local enterprises, the study offers a structured reflection and systematic analysis of how AI can be strategically implemented in educational and creative processes [15]. The conclusions highlight the three key areas: the definition and modelling of an educational project that merges research and design; the identification of strengths and weaknesses in using AI within culturally based fashion design projects; and the critical points to focus on for the attention of future developments of such projects.

The key accomplishment of the project is the development of a replicable educational model that effectively combines research and design practices [16]. It is based on a collaborative model which is grounded in multidisciplinary approach, and which can be adaptable to different local contexts, especially through partnerships with regional fashion and textile companies. This methodology encourages students to explore local heritages as a design resource while with help of AI technologies integrating through the creative processes. This methodology promotes creative innovation and culturally sensitive topics by combining ethnographic research, visual analysis and digital experimentation. The model supports both creative innovation and cultural sensitivity. The adaptability of flexible structure allows it to be implemented across a wide range of geographical and cultural context settings, encouraging fostering meaningful engagement between emerging designers and different territories they inhibit or explore.

The research was able to identify the significant advantages of AI's contribution to heritage-based fashion projects. The research identified several important strengths. First, AI proved to be a very valuable tool in the research phase, particularly in identifying and visually representing personas- archetypal figures that represented cultural identities or values [17]. The AI algorithm systems were able to aid in the creation of mood boards and cultural profiles that are guided in the design phase, through data analysis and image synthesis. Secondly, AI contributed significantly to the initial generation of creative ideas, offering a very fast and diverse set of visual suggestions from which students were able to draw inspiration. Third, it demonstrated considerable efficiency in the visualization of project environments, helping students to contextualize their designs within imagined or in the real-world settings. Finally, the AI-generated outputs provided a very useful starting point for human made graphical elaboration, which remains essential for refining and finalizing the designs. The synergy between digital and manual processes was seen as particularly effective in the project.

However, the research was also able to highlight the critical flaws in use of AI in heritage- related design. One of the major issues, was the tendency to flatten or trivialize cultural references, which often translate into globalized or western-centric representation which diminish the local specifics. In many cases, AI-generated outputs frequently rely on highly simplified symbolic systems which in return fail to capture the depth and richness of traditional imagery. Additionally, the visuals produced were sometimes too ornate and verbose, lacking the minimalism and clarity required in contemporary fashion communication.

There were several key points that were identified by the conducted study to address these issues that designers and educators should consider while working with AI in this context. Firstly, the choice of reference images plays a crucial role: using well-curated, high-quality, and culturally meaningful visual material enhances the aesthetic and conceptual depth of the outcomes and helps avoid stereotypical representations. Second, the precision of prompt writing is crucial. Effective prompts should clearly define aspects such as color palettes, compositional grids, and key symbolic elements, enabling the AI to generate content that aligns with the project's goals. Third, thoughtful choice of environments for contextualizing the design is important to ensure coherence between the garment and its visual setting, careful consideration of the locations to contextualize the design is crucial.

In conclusion, we can find that the integration of AI into fashion projects that focus on local heritage offers both interesting opportunities and a significant number of challenges. Positioned within the framework of regional sustainable innovation, this project offers a replicable model for how design education can activate local ecosystems of creativity and cultural valorization. Through the lens of the Helix model, the course illustrates how academic initiatives can intersect meaningfully with industry and territorial assets to foster responsible, context-sensitive innovation.

By engaging students with tools of advanced digital design while anchoring their work in local heritage, the initiative demonstrates a scalable strategy for integrating technological exploration with regional development goals [18]. This reinforces the role of design not only as a creative discipline, but as a strategic infrastructure for innovation ecosystems that prioritize sustainability, cultural resilience and cross-sector collaboration.

When the approach is critical and creative, AI can support research, spark innovation, and enhance visual storytelling. However, we can understand that AI requires careful curation, cultural awareness, and the constant presence of human interpreting to achieve the result. This study advocated the need of a balanced, ethically informed AI use in fashion education- one that will respect the cultural depth while embracing technological exploration.

Studies indicate that up to 80% of a product's environmental impact is determined during the design phase [19], challenging the common perception of design as merely aesthetic and underscoring its strategic role in sustainable innovation—combining material efficiency, economic value, and scalability potential. This role is further amplified through international mechanisms such as the WIPO Hague System, which enables the global protection and dissemination of sustainable design solutions [20], aligning design practice with the Sustainable Development Goals (SDGs). Moreover, recent research shows that integrating AI into the creative process can further enhance this innovative capacity, provided it is guided by cultural curation and critical reflection [13].

Acknowledgements We thank Professor Arturo Dell'Acqua Bellavitis for his tireless energy and vision in the field of design and fashion, the student Anju Konikkara George for providing valuable support in the writing of this paper. Article written from a paper presented at FIAR 2025, the International Congress of Fashion and Textile Design—FIAR 2025—at IPCB, Portugal.

References

1. Dhaenens F, Esposito J, Willie A (2025) Enhancing loyalty in luxury fashion with AI: the mediation role of customer relationships. Accessed from https://www.researchgate.net/publication/388723118_Enhancing_loyalty_in_luxury_fashion_with_AI_the_mediation_role_of_customer_relationships
2. Gupta A (2024) The rise of AI in the fashion industry. In: Paper presented at the metaverse, fashion design and sustainable fashion conference. Accessed from https://www.researchgate.net/publication/385242729_The_Rise_of_AI_In_the_Fashion_Industry
3. Evangelista PN (2019) Artificial intelligence in fashion: how consumers and the fashion system are being impacted by AI-powered technologies; Politecnico di Milano. Accessed from https://www.politesi.polimi.it/handle/10589/167521
4. CIOReview (2019) How Important AI is in the Textile Industry? Accessed from https://www.cioreview.com/news/how-important-ai-is-in-textile-industry-nid-29709-cid-175.html
5. Amed I, Balchandani A, Berg A., Hedrich S, Poojara S, Rölkens F (2019) The State of Fashion 2020. Accessed from https://www.mckinsey.com/industries/retail/our-insights/state-of-fashion

6. Pasta S (2023) Tra Pedagogia algoritmica e Infocrazia. Recensione a C. Panciroli—P.C. Rivoltella, Pedagogia algoritmica. Per una riflessione educativa sull'Intelligenza Artificiale, Accessed from https://publicatt.unicatt.it/handle/10807/280517
7. Dhaenens F, Esposito J, Willie A (2025) How artificial intelligence and generative AI is revolutionizing the fashion industry. Accessed from https://www.researchgate.net/publication/388793297_How_Artificial_Intelligence_and_Generative_AI_Is_Revolutionizing_the_Fashion_Industry
8. Arora N, Anant BC, Chaudhary K (2023) Optimizing supply chains in the fashion and textile industry through AI: AI and the fashion industry. In: Paper presented at the international conference on fashion apparel and textile (INCFAT23). Amity University Noida. Accessed from https://www.researchgate.net/publication/381109666_Optimizing_Supply_Chains_in_the_Fashion_Textile_Industry_through_AI_AI_and_the_Fashion_Industry
9. Dhaenens F, Esposito J, Willie A (2025) The effects of AI-powered personalization on consumer behavior in fashion e-commerce. Accessed from https://www.researchgate.net/publication/388190363_The_Effects_of_AI-Powered_Personalization_on_Consumer_Behavior_in_Fashion_E-Commerce
10. Mishra MK, Pal R, Nayak R (2025). Chatbots and AI in fashion industry. Accessed from https://www.researchgate.net/publication/389336801_Chatbots_and_AI_in_Fashion_Industry
11. Pasupuleti MK (2024) Fashion in the digital age: AI holography, virtual aesthetics, and the transformation of cultural identity. In: Paper presented at the metaverse, fashion design and sustainable fashion conference. Accessed from https://www.researchgate.net/publication/386324080_Fashion_in_the_Digital_Age_AIHolography_Virtual_Aesthetics_and_the_Transformation_of_Cultural_Identity
12. O'Connel MI (2025) The role of Open AI in fashion: transforming creativity through innovation. Accessed from https://www.researchgate.net/publication/384197232_The_Role_of_Open_AI_in_Fashion_Transforming_Creativity_Through_Innovation
13. Orwig W, Bellaiche L, Spooner S, Vo A, Baig Z, Ragnhildstveit A et al (2024) Using AI to generate visual art: do individual differences in creativity predict AI-assisted art quality? https://doi.org/10.31234/osf.io/ygzw6
14. Franzo P., Cianfanelli E, Quartu A (2024) Co-designing fashion with AI in the phygital scenario. Convergences—J Res Arts Educ 17(34):79–90. https://doi.org/10.53681/c1514225187514391s.34.269
15. Panciroli C, Rivoltella PC (2023) Pedagogia algoritmica: Per una riflessione educativa sull'intelligenza artificiale. Scholé, Milano
16. Suh S (2025) AI technology integrated education model for empowering fashion design ideation. Accessed from https://www.researchgate.net/publication/383447764_AI_Technology_Integrated_Education_Model_for_Empowering_Fashion_Design_Ideation
17. Sbai O, Elhoseiny M, Bordes A, LeCun Y, Couprie C (2018) DesIGN: design inspiration from generative networks. https://doi.org/10.48550/arXiv.1804.00921
18. Ranieri M, Cuomo S, Biagini G (2024) Scuola e intelligenza artificiale: Percorsi di alfabetizzazione critica. Roma: Carocci Editore. ISBN: 9788874669752
19. IISD (2021) The contribution of design to sustainable development. Accessed from https://sdg.iisd.org/commentary/guest-articles/the-contribution-of-design-to-sustainable-development/
20. WIPO (2024) The hague system for the international registration of industrial designs. Accessed from https://www.wipo.int/en/web/hague-system/

Design as a Strategic Factor in Creating Sustainable Musical Instrument Solutions

Ermanno Aparo, Liliana Soares, and Rúben J. Couto

Abstract This paper analyzes the role of material design research to support and develop lutherie production activities, related to principles of sustainability, including, environmental sustainability, social sustainability, aesthetic sustainability and economic sustainability. The paper establishes the importance of the creative process that link design to lutherie, a manufacture area described by its singularity. A productive culture that, over time, has maintained processes, materials and techniques untouchable and unalterable. The paper demonstrates the importance of experimenting and carrying out laboratory tests with alternative and sustainable materials. Methodologically, the study is exploratory and generative. The research addresses a transversal approach, using mixed methods with qualitative and quantitative basis. The exploratory phase was based on a comparative analysis between the performance of some wood senses and some composites that combine a vegetal resin with natural fibers. The woods were chosen according to the four sustainability principles mentioned above. The paper highlights the value of the alliance between design, materials and lutherie as a shared learning mechanism.

Keywords Design research · Design practice · Lutherie · Sustainability

E. Aparo (✉) · L. Soares
Research Centre for Architecture, Urbanism and Design, CIAUD, Lisbon School of Architecture, Universidade de Lisboa, Polo Universitário Do Alto da Ajuda, Lisboa, Portugal
e-mail: aparo@estg.ipvc.pt

E. Aparo
Instituto Politécnico de Viana Do Castelo, Rua Escola Industrial E Comercial Nun'Álvares, Viana Do Castelo, Portugal

R. J. Couto
ProMetheus, Research Unit On Materials, Energy, and Environment for Sustainability Project & Instituto Politécnico de Viana Do Castelo, Rua Escola Industrial E Comercial Nun'Álvares, Viana Do Castelo, Portugal

D. Raposo et al. (eds.), *Strategic Design for Sustainable Futures*, Springer Series in Design and Innovation 65, https://doi.org/10.1007/978-3-032-33282-0_12

1 Introduction

1.1 The Basis

Lutherie is one of the activities that more expresses tradition and production strongly related to local cultures and making needs, such as, native raw materials which are transformed into peculiarities. Consistently, the relationship between the territory and production is related to the availability of the raw material that can allow the creation of the products. This principle is crucial in violin making and maintenance.

For instance, when one talks about the violin forest—Paneveggio—Val di Fiemme and the relationship with Cremona luthier school and, with all the production of violins in neighboring areas, due to the great quality of the spruce, one discus, probably, one of the best materials for making sound box of stringed instruments [1].

Over the centuries, due to natural or human causes, the scarcity of raw materials has put the availability of suitable raw materials to produce musical instruments at serious risks. For example, between 2018 and 2023, the forest of Val di Fiemme lost much of its arboreal heritage, due to an insect (Ips typographus) and a violent storm (the Vaia), putting the availability of the 'spruce [2].

1.2 Related Work

There are some case studies that endorse this study.

In the United States, for instance, there is the spruce used for the musical instruments of brands such as Martin and Gibson. The extreme exploitation of the forests of Appalachia was causing the deforestation of the territory and led the American government institutions to a protectionist policy, preventing and protecting not only, the remaining forest heritage, but also the geological conditions of the areas affected [3]. These situations have also succeeded in other countries, such as, the Scandinavian ones, Switzerland, France or in the Carpathian Mountains in Romania. Climate change policies have been implemented to limit the abatement of trees to a few months of the seasons winter [4].

The search of high acoustic performance or the maintenance of traditional materials always had to refer to the difficult availability of some good species, making it very often necessary to search for alternative solutions to traditional ones. The creation of instruments with local raw materials is something common and ancestral. In ancient Greece, the Kithara and the Lyra—as archetypes of modern stringed instruments—were built referring to the types of wood available. This is the case of boxwood (Buxus sempervirens) which, in the Peloponnese areas, was used for the construction of Lyras to replace turtle shells and animal horns [5].

Another case that can be highlighted is the cedar—or cypress—considered types of sacred wood and, which could be used for sacred purposes such as the creation of instruments such as the Kithara [6].

Gasparo Bertolotti da Salò (1540–1609), an Italian luthier and direct opponent of Andrea Amati in the construction of stringed instruments, used essences from the Pinaceae family such as cedar, justifying his choice with the availability and quality of the material. Cedar was very widespread in local construction, guaranteed easy availability and a regular distribution of the grain. These characteristics assured high sound qualities, like those ensured by spruce and maple, normally used for sound tables of stringed instruments [7].

Throughout the ages it was created a connection around luthiers' image and legacy. The production and processes have been highly romanticized, creating a mysticism that justify the continuous research produced by craftsmen. A case that underlines this line is Giuseppe Guarneri del Gesù as, it seems, he developed the "the servant's violins" while he was in prison. It, also, seems that it was a servant who provided him with wood, in the prison, to make some of his instruments. This enquiry optimized the material income and the search for valid and performative options [8].

The research in string musical instruments led the researcher and luthier, Carleen Maley Hutchins, to determine that the high transversal rigidity of the grain was a central factor to create wood quality in the creation of harmonic piano. Hutchins consolidated this finding during the creation of the soundboard, with other types of wood such as pear, apple, cherry and teak [9].

This investigation and, consequently, the development of materials science, have led to the concrete possibility of replacing traditional materials with synthetic alternatives which, in various parts of the world, have been used to change natural wood. For instance, these are the cases of carbon fiber violins from the German company "Mezzo Forte" [10] and the North American brand "Clark" [11].

Today, research provides materials that can be produced with high precision, allowing the creation of components less susceptible to deformations or changes due to humidity and temperature. Furthermore, synthetic materials often have a lower cost, theoretically increasing access to quality instruments for a wider range of musicians. However, the use of synthetic materials raises concerns about the potential loss of authenticity and tradition in luthier craftsmanship. The stereotyped use of artificial materials can lead to a dangerous and flattering standardization on the values of uniqueness that have always distinguished this art. The relationship created by the luthier with a natural material such as wood participates in the creation of a singular sound and unique enjoyment for the musician.

The lack of this individuality can cause a loss of quality of the product but also of those who use it and choose their tools. In this sense come the experiences made by some designers and luthiers who, following principles such as those of social and environmental sustainability, decide to develop projects using natural fibers and guaranteeing compatibility with the production culture of musical instrument construction.

For instance, designer Luca Alessandrini, in cooperation of Cremonese luthiers, developed some violin prototypes with a natural fiber using spider web [12]. On the other hand, Tim Duerinck, luthier and researcher at the University of Ghent used linen fiber. These examples explain the strong link that had been created between the luthier production culture and the territory. This connection supported processes and

Table 1 Woods' choice according to 4 principles of sustainability

Who	What	Where
The authors	Species of wood normally used by local companies or artisan's workshops for the construction of stringed instruments The availability of autochthonous species to be purchased and acquired	North of Portugal

The Authors (2025)

structured knowledge that have been transformed into heritage. In this sense, design can contribute to the construction of alternative ways to manufacture violins. Sustainable paths that do not replace traditional paths and that, for this reason, strengthen and improve the productive capabilities of the territory [13].

2 Methodology

The methodology used for this research was exploratory and generative. Based on a transversal approach, it used mixed methods with qualitative and quantitative basis [14, 15]. The exploratory phase comprises a comparative analysis between the behavior of some wood senses and some composites that combine a vegetal resin with natural fibers. The woods were chosen according to principles of sustainability, such as, environmental sustainability, social, sustainability and economic sustainability (Table 1).

3 Research Process

The experience of some luthiers in the region described the spruce (picea abies), maple (acer) and lime (tilia) as main factors provided by luthiers and acquired from distributors. During the process it was decided not to carry out acoustic tests. According to several authors, such as Carleen Hutchins [16], acoustics can be influenced by the shapes and morphology of the components of the musical instrument. To make ideas concrete and tangible, the sooner it will be able to evaluate them, refine them, and zero in on the best solution [17] (Fig. 1).

3.1 Mechanical Tests with Wooden Specimens

The first phase consisted of carrying out mechanical tests with wooden specimens. Thus, different types of wood were used, such as green spruce, talia, maple, hemp, flax, and flax hemp. Some mechanical tests were carried out to collect data as a reference for the composite material that will be analyzed (Fig. 2).

Fig. 1 Boards of wood species locally used for the construction of stringed instruments. The Authors (2023)

Fig. 2 Wooden specimens sized for mechanical tests. The Authors (2023)

In this phase, it was decided to spread by analyzing tensile tests, which are important for testing the dynamic behavior of materials in musical instruments (Fig. 3).

The entire process had the support of computational design and generative design models and tools to reduce traditionally time-consuming processes, expose new combinations and helping the phase of directing tests and compiling results [18].

The tests also had the objective of analyzing Young's modulus, a property of the material that express how it can stretch and deform and is defined as the ration of

Fig. 3 Materials and models needed to carry out the tests. The Authors (2023)

tensile stress to tensile strain [19]. Thus, it is an important factor to understand the resonance of the material.

To homogenize the tests and make the analysis more objective, the tests were performed, using specimens with identical geometries (length, width and thickness).

To carry out the analyses, the tests resulted in models to be used in mechanical tests with fiber and resin specimens.

3.2 Mechanical Tests with Fiber and Resin Specimens

For mechanical tests with fiber and resin specimens, it was chosen different materials and binders.

The first experiment was prepared with Rosin Resin, which comes from the processing of pinnace or conifer resins and, which is also used to improve the friction of the bow on the strings of bowed musical instruments.

The second test was done with an Eco-Resin produced with a base of natural and vegetable products with performances and preparatory processes very similar to those of classic epoxy resins.

The third assay was done with Hemp Fiber, reproduced in Portugal, currently considered one of the organic fabrics of the future. After, it was done an experiment with flax fiber, a material frequently produced in Portugal, that presents excellent performance on a mechanical and structural level (Fig. 4 and Table 2).

The tests were carried out by creating specimens obtained by combining the resins and fibers together, in such a way, to be clear which of the possible combinations could be most suitable for the first phase, and possibly, for the others. Another factor was the comparison with a composite of epoxy resin and carbon fiber, as it is widely used in the production of musical instruments as an alternative to wood (Fig. 5).

Fig. 4 Fiber and resin specimens sized for mechanical tests. The Authors (2023)

Table 2 Tests

Materials	Location/Process	Relation with music
Rosin resin	Comes from the processing of pinnace or conifer resins	To improve the friction of the bow on the strings of bowed musical instruments
Eco-resin	A base of natural and vegetable products existed in the place	Performance like that used in classic epoxy resins
Hemp fiber	Produced in the place	Organic fabrics of the future
Flax fifer	Produced in the place	Excellent performance on mechanical and structural level

The Authors (2025)

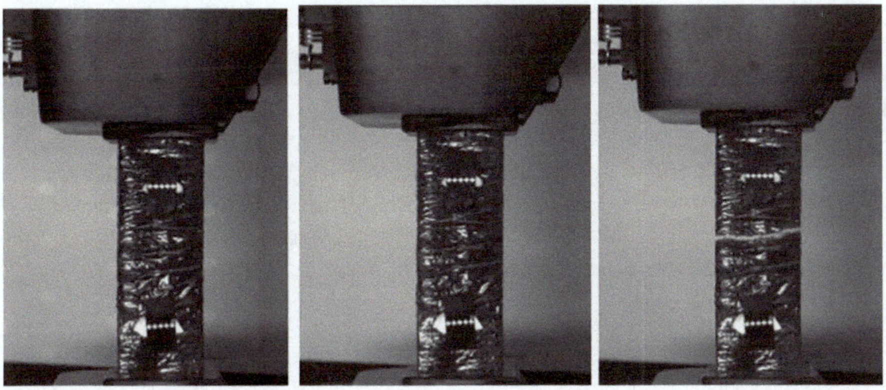

Fig. 5 Fiber and resin samples tested on a universal woodworking machine. The authors (2023)

The creation of a new composite cannot just respond to acoustic or mechanical needs but must necessarily satisfy other needs that can guarantee the reproducibility of the material and the processing by the luthiers. In this sense, it seemed important to analyze processes that could be related with the luthier's work.

These tests have directly to do with the creation these factors, but some shapes were abandoned because, depending on the type of material and the type of shape to be developed, they can be subject to significant alterations one from another.

In this sense, it was made some progress with the preparation of some test pieces, depending on the study of previous mechanical analyses.

3.3 Tests and Results with Rosin Resin Specimens

The analysis continued with Rosin Resin specimens, but the test preparation demonstrated the ineffectiveness of Rosin Resin option, due to the different considerations, such as:

1. The speedy solidification was very fast and difficult to control, inhibiting a correct and homogeneous distribution of the material in the mold.
2. The rosin procedures temperatures were very low, compromising several phases of the process and producing formal changes, in tune with proper use of body heat.
3. The process requires a laboratory structure with controlled equipment, quite different from those, normally, available in a homemade workshop.

The specimens were subjected to tensile testing using the Shimadzu AG–X plus machine. Flat grips were used to secure the specimen. Two black stickers with white diamond patterns were applied. In this way, the software, together with the camera, can track the position of the two markers, functioning as a visual extensometer by calculating their displacement as the test progresses. In a tensile test, the lower grip remains stationary while the upper grip moves upward until the specimen fractures. The force (N) and displacement between the markers (mm) are recorded (Figs. 6 and 7).

This phase also highlighted that the tests produced with Rosin Resin were not homogeneous and were extremely fragile. This fact determined that its use would be irrelevant in the mechanical tests during the second phase. Consequently, the mechanical samples tests are of three types (Table 3).

4 Results

The thicknesses of the wooden samples were double those of the resin and fiber samples and that the composites were not produced with optimization processes to avoid inhomogeneity between the surfaces, we can conclude that the results obtained

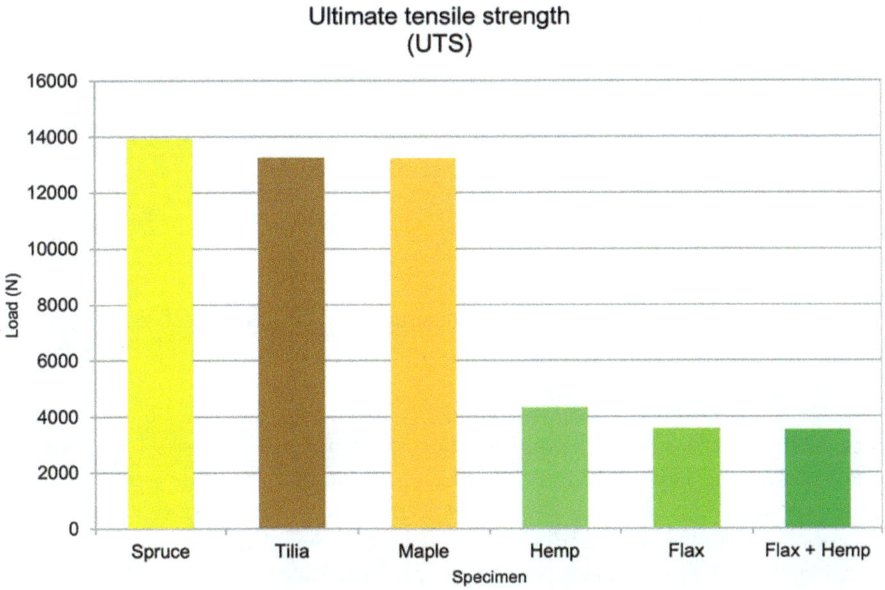

Fig. 6 Comparison of the tensile strength of: Spruce, Tilia, Maple, Hemp, Flax, and Flax Hemp. *Source* The Authors

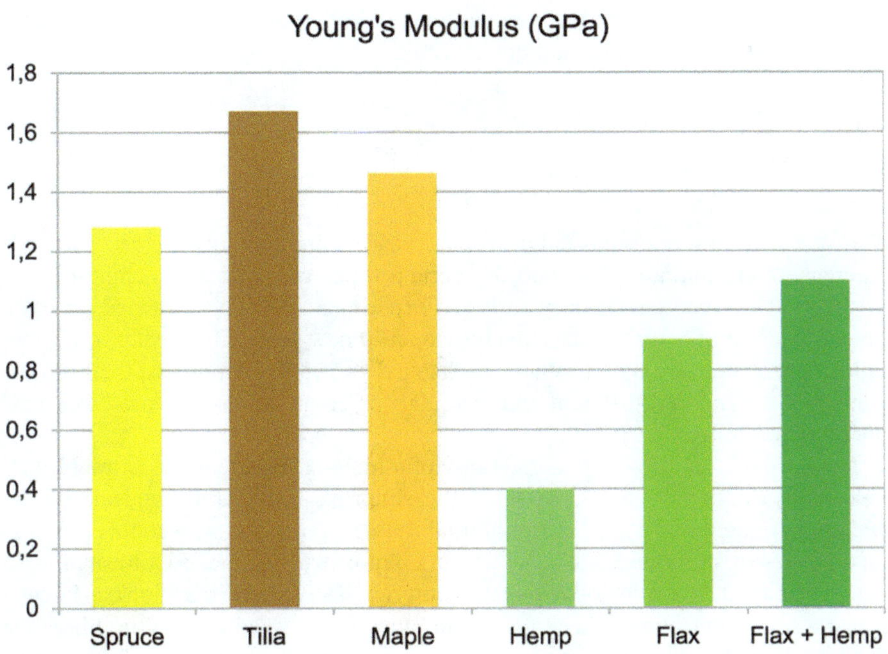

Fig. 7 Comparison of Young's Modulus of: Green Spruce, Tilia, Maple, Hemp, Flax, and Flax Hemp. *Source* The Authors

Table 3 Mechanical sample tests

Materials	Mechanical test types	Results
Rosin Resin	Sample with linen fiber and eco-resin	Tests were not homogeneous
	Sample with hemp fiber and eco-resin	
	Test body with the two fibers placed alternately and eco-resin	

The Authors (2025)

allowed us to determine that flax fiber and hemp fiber, combined with eco-resin, can give rise to valid alternatives for the creation of a composite capable of replacing the traditional offer of wood essences such as spruce or maple.

Unfortunately, in this phase, the laboratory conditions did not allow to consider all the factors necessary for the optimization of the mechanical tests. Therefore, the results, particularly those relating to the calculation of the tensile strength, probably, could have been better than those obtained in this phase. Considering that the composite currently most used as an alternative to wood as a material is carbon fiber, which both as raw materials and as processes raises serious doubts about its sustainability and the process management capabilities of a luthier, we can conclude that the tests produced can be a credible basis for the creation of a more sustainable alternative to this type of material.

The results contribute to the constant analysis of designing products with high efficiency and low environmental impact [20].

5 Conclusions

Over time, the culture of musical instruments and violin making have demonstrated coherence and continuity by choosing certain types of materials, techniques and processes. Throughout its history, tests and experiments have been carried out, with the introduction of new materials that have opened new production avenues, without replacing tradition. Both parties have exchanged knowledge, improved their professional skills, proposed different and spontaneous connotations and produced new market opportunities [21].

The introduction of new materials in violin making represents an extraordinary opportunity to expand expressive possibilities, improve sustainability and address the challenges related to the availability of traditional resources. However, this progress must not result in a break with the artisan culture that has made violin making a recognized art and handed down over the centuries. The lutherie tradition can be seen as just a set of consolidated techniques but must be analyzed as a cultural heritage in which knowledge, sensitivity and mastery converge to give life to instruments with a unique soul. New materials must therefore fit into this context with respect and caution, adapting to consolidated artisan processes and preserving the aesthetic, acoustic and constructive value of the instruments, but also being aware that in violin making a process can be transformed into a unique ritual and that this also contributes to giving an identity to a musical instrument. Rather than indiscriminately replacing wood and other historical materials, innovation must create compatible alternatives that enrich the violin maker's repertoire without distorting its identity. Research on innovative materials, in fact, should be guided by a deep understanding of tradition, so that new solutions respect the sound, tactile and visual needs of classical instruments. The evolution of lutherie must therefore be understood as a dialogue between past and future: only through a careful study of tradition can a valid alternative arise that should not claim to replace tradition but to accompany it to ensure its survival.

Design of a musical instrument can be seen as a creative challenge and an opportunity to suggest the use of unconventional natural materials in its production. From an aesthetic point of view, sustainable design should aim to create long-lasting objects that do not camouflage themselves with tradition but can guarantee their own identity, ensuring that the user can discern tradition from innovation. At the same time, the use of new sustainable materials must offer musical instruments the possibility of being attractive enough (aesthetically and functionally) to encourage users to use and keep them [22]. In a modern vision, aesthetics combines functionality and sustainability in the creation of a musical instrument that becomes unique in its existence, which includes its appearance and sound. A complex and creative process that involves experimentation and different fields, and that uses lutherie as a complex phase for an innovation path, which can be subsequently transferred to other fields or areas of production and knowledge.

Acknowledgements Ermanno Aparo and Liliana Soares acknowledge CIAUD—CIAUD, Research Centre for Architecture, Urbanism and Design, Lisbon School of Architecture, Universidade de Lisboa, Lisboa, Portugal. Rúben J. Couto acknowledge proMetheus—Research Unit on Materials, Energy, and Environment for Sustainability project, FCT Ref. UID/05975/2020, financed by national funds through the FCT/MCTES.

References

1. Fratus T (2016) L'Italia è un bosco. Laterza, Bari
2. Favero P, Carniel S (2019) C'era una volta il bosco: gli alberi raccontano il cambiamento climatico: sarà una pianta a salvarci? Hoepli, Milano
3. Waugh-Quasebarth J (2019) Finding the singing spruce: craft labor, global forests, and musical instrument makers in Appalachia. University of Kentucky, Lexington
4. Whited T, Engels JI, Verstegen W, Ibsen H, Hoffmann RC (2005) Northern Europe. Bloomsbury Publishing, London
5. Ciervo C (2019) I cordofoni nell'antica Grecia: una riflessione tra mito e tecnica. In: Annali 2019. Suor Orsola Benincasa Università Editrice, Napoli, pp 57–87
6. Bottini A (2021) Salvezza e archeologia: il caso di metaponto. In: PIZZIRANI, Claudia (Cur) Iconografia e rituale funerario Atti del I° Incontro di Studi sul significato delle immagini nei contesti funerari. Alam Mater Studiorum, Bologna, pp 9–36
7. Ravasio U (2012) Liuti, violini, lyre e simili. Archivio di Stato, Brescia.

8. Graziosi G, Graziosi S (2020) Il violino e la sua grande famiglia. Bibliotheka edizioni, Roma
9. Hutchins CM (1981) L'acustica dei piani armonici di violino. In: Le Scienze, pp 86–99
10. LNCS Homepage, https://www.mezzo-forte.de/gb/22-carbon-fiber-violin, Last accessed 03 Dec 2024
11. LNCS Homepage, https://luisandclark.com/the-instruments/the-violin/, Last accessed 03 Dec 2024
12. LNCS Homepage, https://lucaalessandrini.com/silk-violin, Last accessed 03 Dec 2024
13. Aparo E, Soares L, Gonçalves E (2023) The will-to-power to design a violin. In: Human dynamics and design for the development of contemporary societies, vol 81, pp 129–138
14. Creswell JW (2009) Research design: qualitative, quantitative, and mixed methods approaches. Sage Publications, London
15. Laurel B (2003) Design research, methods and perspectives. The MIT Press, Cambridge
16. Hutchins CM (1995) Measurable characteristics of violin-family instruments in relation to the sound of a high-quality violin. MRS Bull 29–31
17. Brown T (2009) Change by design: how design thinking transforms organizations and inspires innovation. Harper Collins, New York
18. Saadi J, Yang M (2023) Observations on the implications of generative design tools on design process and designer behaviour. In: Design society, N° 3, pp 2805–2814
19. LNCS Homepage, https://www.sciencedirect.com, Last accessed 25 Jan 2025
20. Soares L, Aparo E, Almendra R (2024) Designing a creative process between music, performing and visual arts and the business world. In: Springer series in design and innovation, vol 35, pp 336–344
21. Soares L, Aparo E, Almendra R (2025) Storytelling in design: in digital age cultural show travels on network. In: Springer series in design and innovation, vol 48, pp 443–453
22. Harper K (2018) Aesthetic sustainability. Product design and sustainable usage. Routledge, London (2018)

Design Management for the Maintenance of Cultural Projects in Brazil

Valéria Ramos Friso, João Carlos Riccó Plácido da Silva, and José Carlos Plácido da Silva

Abstract Making a living from art is a challenge and a significant achievement for any artist. While in the past the aristocracy was the main funder of artistic productions, today there is greater democratization of access, driven by public policies (although still limited) and direct support from the public. Planning has always been essential in administration, especially according to Neoclassical Theory. In this context, design emerges as a valuable tool for planning in the artistic field, evolving from a discipline focused on physical objects to a broader and more strategic approach. Strategic Design acts as a planning method that defines the interface between companies, the market and society, stimulating organizational creativity and fostering innovation through the exchange of knowledge. The integration of these processes allows cultural projects to be guided with design management methods, seeking greater sustainability and efficiency in artistic production.

Keywords Management · Design · Cultural projects

1 Current Context

Working exclusively with art represents a challenge for professionals in the field today. If in the past, aristocrats were the main consumers and financiers of classical art, today we find initiatives to democratize access, based on public policies (albeit insufficient) and alternatives for direct financing by the public. While the figure of the aristocrat or the government can be identified as a unit, the public is made

V. R. Friso · J. C. R. Plácido da Silva (✉)
UFU—University Federal of Uberlandia, Uberlandia, Brazil
e-mail: joao.placido@ufu.br

V. R. Friso
e-mail: valeriafriso@yahoo.com.br

J. C. Plácido da Silva
UNESP—São Paulo State University, Bauru, Brazil
e-mail: jcplacido@gmail.com

D. Raposo et al. (eds.), *Strategic Design for Sustainable Futures*, Springer Series in Design and Innovation 65, https://doi.org/10.1007/978-3-032-33282-0_13

up of heterogeneous groups. To improve their income, groups are organized into organizational structures that can respond to the demands of potential subsidiaries. There are differences, however, in the way these groups are constituted in the different areas of the cultural market, which is a result of values traditionally cultivated in each segment, consequently generating different results in terms of productivity.

The cultural sector in Brazil employed 4.8 million workers in 2020, according to the Institute of Applied Economic Research (Ipea), but a significant portion of these professionals work without a formal contract or in temporary jobs. According to the same study, approximately 44.9% of the sector workers in the sector are self-employed or informal workers. This reality makes them. Faced with economic challenges, artists from different areas are mobilizing, through collective and individual initiatives, to ensure that their work is not interrupted, both by financial difficulties and by blocking the inspirational process of their activities. In the meantime, thinking about strategic actions that consider and reconfigure the context can generate new resources, or improve existing ones, which can be used to benefit creative activity.

Planning has always been one of the pillars of management, especially in the context of the Neoclassical Theory of Management, which considers it one of the stages of the administrative process. According to this theory, the four basic administrative process functions are: planning, organization, direction and control. According to Chiavenato [1], planning consists in establishing objectives and choosing the means to achieve them; organizing involves allocating resources and people to achieve the established objectives; direction is the process of guiding people in the development of goals; and control refers to monitoring activities to ensure that the organization remains informal and achieves the objectives, making adjustments when necessary.

The concept of strategy emerged in the post-war period in the 1950s. The term is broad and can take on different interpretations depending on the context in which organizations or individuals find themselves [2]. In the business world, Porter [3] defines strategy as the combination of a company's objectives and the means necessary to achieve them. Chiavenato and Sapiro [4] present organizational strategy as the way in which an organization can compete, survive and remain sustainable over time, considering social, cultural, financial, economic and technological aspects. Although there are several approaches, the main purpose of organizational strategy is to ensure that the organization achieves its objectives and goals [5]. Understanding strategy is seen as a social practice and a complex and constantly evolving phenomenon [6].

Technology plays a crucial role in developing strategies. The Fourth Industrial Revolution, or Industry 4.0, has brought about a series of technological changes that are transforming production, consumption patterns, and the functioning of societies in general. The Fourth Revolution is characterized by the digital integration of physical and virtual systems, with the intensive use of technologies such as artificial intelligence (AI), the Internet of Things (IoT), big data, advanced robotics, 3D printing, nanotechnology, among others. In practice, the integration of physical and digital technologies has influenced the way artistic works are produced. Communication between devices, with broader systems, favors personalization and proximity between those involved in projects.

The ability to collect and analyze large volumes of data allows more informed decision-making and more agile and adaptive strategies. Artificial Intelligence favors diagnostics, process optimization, and large-scale data analysis, saving time with low investment, becoming an important tool for emerging groups. Technology is intrinsically associated with innovations that promote economic and social transformations, influencing the processes of production and distribution of goods and services.

Design, as a mediator between planning and technology, has made important contributions to the art field production. Over time, design has evolved from a discipline focused on the design of physical objects to a broader field that encompasses the creation of solutions for a wide range of problems in different contexts. The designer seeks to make the user experience increasingly satisfactory. It includes everything from the visual and sensorial world to the adequate management of resources for sustainability [7].

This study establishes a connection between design management and the maintenance of cultural projects, demonstrating how this relationship can promote more sustainable processes. To this end, it conducts a bibliographic survey on the concepts of cultural projects, design methodologies, and sustainability, supporting the development of methodological guidelines for continuous and economically viable cultural productions.

2 Economic Sustainability in the Arts Market in Brazil

Economic sustainability refers to the sector's ability to remain financially viable, ensuring, in the case of arts ventures, the livelihood of artists, the stability of cultural institutions, and the continued public access to works. In Brazil, this issue involves challenges such as the dependence on public incentives, the informality of the sector, and the concentration of opportunities in large urban centers.

A study by the Brazilian Institute of Geography and Statistics (IBGE) revealed that, between 2011 and 2021, the number of companies in the cultural sector grew by 3.1%, totaling 387.5 thousand in 2021. However, the share of these companies in the national total decreased from 7.3 to 6.7% in the same period. In Brazil, the Special Secretariat reported in 2020 a variation in the sector's share of GDP between 1.2% and 2.67% (Special Secretariat of Culture 2020). The importance of a close look at the cultural market in Brazil is undeniable, not only in the analysis of its results, but also regarding financing initiatives.

Historically, the tension that exists between art and the market is notable [8] in "The Privatization of Culture" problematizes the way in which art began to relate to economy, especially throughout the 1980 and 1990s. For the author, government and corporate interests instrumentalize culture as a "means of social distinction", a "propaganda strategy", and a "customer acquisition", standardizing the values and tastes of those belonging to the same social groups.

It is common sense that the nobility of art is inversely proportional to its commercial appeal. It is sometimes understood that the commercial nature reduces the quality of works of art, as if there were a sacredness that money deprives. According to Benjamin [9], art has always been linked to ritual and cult functions, derived from its connection with magic and religion. As works of art can be distributed on a larger scale, which can affect the authenticity criterion. On the other hand, the possibility of reaching the masses arises, and a change in the art perception and meaning by the public. Thus, it is important to consider the level of of proximity and familiarity with the artistic object, considering the relational condition of art (sender/receiver).

The French economist Xavier Greffe states that the commonly suggested distance between art and the market arises from the fear of artists that economic logic will start to direct their artistic activities [10]. Thompson [11] argues that the monetary devaluation of art lies between the work of art and exchange relations, so that the commercial value of a work of art never depends exclusively on its "artistic quality". Changing the status of the value attributed to the service or product generates economic instability related to the art market [12].

Regardless of value relations, art, like any other product, requires resources. Art can be promoted by its own investment, by external investors, or even by government incentives, with the latter being the main founding source for small companies and experimental groups. However, cultural policy researcher Marila Vellozo (2015) discusses how public notices are ineffective in providing and encouraging "spectacle that require temporary consolidation, such as the maintenance of permanent artistic groups and companies" [13]. Barbosa [14] corroborates this statement, arguing that there is a lack of political articulation to structure and provide continuity, permanence, and sustainability for artists.

According to the IBGE, the number of municipalities with specific resources for culture fell from 90.5% in 2014 to 77.8% in 2021. According to the Continuous National Household Sample Survey (PNAD), only 27% of the professionals in the cultural sector were able to access some kind of tax incentive or financing in 2021. The unequal distribution of investments also harms artists and cultural entrepreneurs from regions outside the South-Southeast axis, which concentrate around 80% of tax incentives for culture. Barbosa also points to the lack of understanding about the need for sustainable cultural policies. [...] The "subject agent condition" is precisely what which arises from a sustainable cultural policy, capable of creating autonomy for the artist subject, making him/her a de facto cultural mediator. The State is responsible for this political-legislative reorganization, to be conducted in harmony with the artists [14]. Although public initiatives are a viable path for independent artists, through the development of strategies, they can find other equally efficient (or more) paths not only for for group's artistic work realization, but also for its own survival.

The success of an organization, in turn, cannot be attributed solely to profit. According to Marion [15], "the recognition of intangible assets has gained importance in the information age, where the value of companies is increasingly related to their capacity for innovation and brand rather than to the physical assets they own." Intangibles are at the heart of artistic work, which presents an opportunity if used well. The study of assets dates back to the beginnings of accounting, being formalized with the double-entry bookkeeping theory, developed by Pacioli [16]. However, the distinction between tangible and intangible assets gained relevance with the evolution of accounting standards and the modern economy. Intangible assets are those that do not have physical substance but generate future economic benefits for the company. They are usually related to legal rights or intellectual property, such as copyright. According to Iudícibus et al. [17], "intangible assets represent resources controlled by the entity that do not have a physical presence but are fundamental to its generation of value." According to Marion [15], "tangible assets are those material goods that can be touched and measured objectively within the company's accounting." The distinction between tangible and intangible assets reflects economic evolution, moving from a model based on physical goods to a model in which knowledge, innovation and technology are the main generators of value.

The art market in Brazil is marked by a duality: on the one hand, there is a growing appreciation of national contemporary art on the global stage; on the other, the sector still faces difficulties in consolidating a sustainable business model. Artists face challenges in a saturated market, with little support from influential galleries, investors or critical attention. When their work becomes viable, they still have difficulty stabilizing their financial situation, surrendering to the need balance the creation of art with other sources of income, such as side jobs. Furthermore, in Brazil, a significant dispareity between centers can be seen such as São Paulo and Rio de Janeiro, and other regions of the country [18].

Strategic planning is important to direct investment in tangible and intangible assets, as well as to guide the very formatting of art ventures in Brazil. The assimilation of new knowledge with the participation of professionals from other areas contributes to the optimization of work and greater profitability for the team, ensuring better results in the medium and long term. The question remains how to make planning viable within the process.

3 The Reality of the Artist

Post-Fordism and the growing consolidation of precarious working and living conditions are often associated with the expansion of immaterial labor. Lazzarato [19] conceptualizes immaterial labor as what generates the informational and cultural content of goods, which is the Artistic activity foundation. In this context, small "productive units," represented by individuals, are mobilized for specific tasks and tend to exist only during the execution of these projects. From this perspective, Boltanski and Chiapello [20] developed their analyses of "the new spirit of capitalism," highlighting the increase of the economy around project-based work estructure. This model promotes the alternation between different initiatives, driving new collaborations and partnerships and favoring the formation of networks.

Lupinacci [21] highlights collaboration as an essential element within the art environment, including professionals responsible for the production of costumes, stage materials, lighting, sound design, and interpretation-creation. The author emphasizes that this collaboration goes beyond the relationships between performers-creators, producers and directors, directly impacting the processes of choreographic composition in contemporary times. In a complementary manner, Luz [22] discusses the sharing of actions in cultural production as an opportunity for young people to develop skills in various areas, such as organizing costumes and props, scheduling new teachers, production and lighting. According to the author, every company is invited to follow certain artistic processes, participating in the development of costumes, scenery and lighting, in addition to assisting in the shows executive production, contributing to technical aspects such as dressing rooms, stage operations and box office. The proposal seeks to provide a comprehensive view of the needs that are inherent to the practice and logistics of artistic dissemination.

Although versatility can be considered a positive characteristic, when imposed by a lack of resources, it results in work overload. Martin [23] was one of the first theorists to discuss the precariousness of artists' work, stating that "precariousness, ephemerality, and instability are often expressed as laments by artists, performers, and audiences. They too struggle to survive; performance venues suffer from dwindling support; audiences face rising ticket prices" [24]. In response to this reality, many artists take on multiple responsibilities, which can lead to physical, mental, and emotional overload. In addition to the risk of burnout, this accumulation of functions can compromise creativity. According to Tenenblat [25], overload results from the "proportionality and creative flexibility of its members", since proposing ideas and actions to the collective requires that each member have the knowledge and ability to articulate their creative intentions. Nascimento [26] reports the testimony of an artist who faces difficulties due to the lack of recognition of artistic activity as work, the absence of formal employment relationships and benefits, financial insecurity, the intense pace of work and restrictions on creative freedom, including teaching dance classes, choosing music, developing choreography, costumes and sets. The author discusses the relationship between artistic work and suffering, arguing that market demands often override the needs of artists, cultural organizations and the public [27]

Although the tension between art and the market has been constant throughout history, financial exploitation in the arts in the twentieth century reached unprecedented levels. Her qualitative research revealed that dancers often feel pressured and overwhelmed by the multiple roles assigned to them. In this sense, art, ideally intended to harmonize the unmet desires of the artist and the public, ends up causing suffering to the creator [26]. Van Assche examines the functioning of the contemporary art sector in Europe, focusing on the dance scenes of Brussels and Berlin. Her research addresses precariousness as a relevant social issue in the context of contemporary artists and highlights the contradictions inherent in the project-based artistic labor market. In addition, the study investigates how these conditions impact the aesthetics and content of the works produced.

Current art investment and consumption relations are the result of decades of conceptual construction, private interests and public policies. In addition to the need for transformations at a macro-structural level, it is essential to rethink the composition of the teams that make up companies, adopting conscious strategies and promoting integration into sustainable development networks. In this way, it becomes possible to provide better working conditions for artists, ensuring the quality of production and fostering collective growth, which, in turn, drives the cultural market.

4 Design for Planning in the Arts

Current art investment and consumption relationships are the result of decades of constructing concepts, private interests and public policies. In addition to fighting for macro-cultural changes, it is possible to rethink the composition of the teams that make up arts companies based on conscious strategies, as well as integrating a sustainable development network to provide art workers with the conditions to produce with quality and prosperity, promoting collective growth and thus fostering the cultural market.

Production design is perhaps the discipline that most clearly brings strategic design closer to the performing arts in terms of organization and planning in the current scenario. At the 85th Academy Awards, the "Best Art Direction" category was changed to "Best Production Design". The Art Director category was renamed the Designer category. ("SPECIAL RULES FOR THE MUSIC AWARDS"). This seems coherent, since the activity is more in the field of planning than directing itself.

Production design is the area of audiovisual and cinema responsible for the visual aesthetics of the film, which includes the organization of sets, furniture, objects and their interactions with the actors [28]. Production design not only deals with the symbolic meanings of images but also determines how suitable the appearence of the scene is to the intended traget audience or market. Regardless of the size or style of the design project, the visual conception implementation conception must consider a specific set of conditions and limitations related to each production. These limitations include issues of cost, technical and human resources, availability of materials and time. The design of a scene must be applied in an integrated manner to the various considerations that move between a rigid dimension and abstract conceptions of meaning construction, considering the objectives of the institution versus the producer. All work must be anchored to an essential condition, which is intrinsic to the cinematographic product: to serve as a vehicle for satisfaction, aesthetics and artistic expression [28].

Design presents itself as an ally of the arts with great potential from various perspectives. Considered by Bomfim [29] to be transdisciplinary, its practice requires knowledge of sciences considered classical and prior to design, such as physiology, mathematics, aesthetics, human sciences, etc. It involves interaction and cooperation between different disciplines, leading them to share a common set of fundamental concepts or the same paradigm, symbolizing unification. According to Pereira [28], a designer, when creating solutions for society in the form of products or systems, must go beyond disciplinary barriers to develop effective responses, from the perspective of transdisciplinary. Each relationship established between the different areas brings to light a new field of interaction and sharing of scientific concepts from both.

The connection between Design and strategy can come from two distinct points: the perspective of administrative culture and project culture. In the field of Administration, experts point to Design as a tool capable of distinguishing products from competitors, providing a competitive advantage for the company [29]. In the face of social and economic transformations, researchers have noticed a new contribution of the activity to organizations: a process capable of boosting organizational development and guiding their strategies (ZURLO, 1999). Rachel Cooper (United Kingdom), Kathryn Best (United Kingdom), Thomas Lockwood (United States) and Brigite Borja de Mozota (France) are some of the researchers who offer valuable concepts on the strategic role of Design in companies, using a perspective of managerial culture in relation to the Design process (COOPER; PRESS, 1995; Best, 2006; DE MOZOTA, 2003; LOCKWOOD, 2010, MAGALHÃES, 1997). On the other hand, the Italian school of Polytechnic di Milano also addresses this role of Design, but from the perspective of project culture. In the same line of thought, Ezio Manzini and Anna Meroni explore Design in the context of organizational strategies focused on Social Innovation.

Zurlo [30] recognizes the benefits of project culture in the formulation of organizational strategies and develops a theory on the strategic role of Design, based on the contributions of other authors. For the researcher, in the modern company, strategy is perceived as "a collective process of construction of meaning". The strategies developed by these organizations allow them to adapt to changes, maintaining and redefining their own identity, while impacting the external environment. In the collective sphere, strategic design activates organizational creativity, promoting new knowledge to drive innovation and exchange. Given that organizations operate in value co-production networks, strategic design mobilizes stakeholders, employees and citizens to co-create value.

Cain [31], in suggesting the methodology "experience-based design", proposes a reorganization of the design process centered on the understanding of companies, their products and services, and the experience of the people who use them. He defines the method as "a process that uses a deep understanding of people's everyday use and experiences with products and services to inform and shape business objectives" [32]. He identifies the components of experience as: (1) sociocultural systems that shape users' ideas, beliefs, attitudes, and expectations (think); (2) patterns of action and routines that form identity and meaning (do); (3) objects that people use and their impact on what they think and do (use).

Meroni [33] clarifies that community-centered design emphasizes understanding social needs and behaviors, rather than individual ones, adopting a co-design attitude to collaborate closely with the community in formulating strategic changes. Similar to experience-based design, community-centered design is grounded in Geertz's interpretive anthropology [32], which observes and describes in detail the layers of meaning networks that are essential to creating narratives of future scenarios. A Design-Driven Scenario is an articulated and motivated vision that aims to mobilize the energy of the various stakeholders in the design process to direct their efforts towards a shared vision. These scenarios are marked by plurality, feasibility, micro-scale, visual expression and participation. Their construction considers alternative contexts where environmental, sociocultural and economic aspects are assessed, based on available technologies or identified sociocultural opportunities. They are defined by the scale of people's life context, visually illustrating contexts and proposals to engage a diverse group in a collective project.

A scenario must contain three elements: vision, motivation and proposal. The vision answers the question "what would the world be like if…?" using a narrative that summarizes the situation after a sequence of events. The narrative demonstrates how the context of an artistic organization can be modified if certain behaviors are adopted, actions taken and product and service proposals implemented. A physical and sociocultural environment can be created in which the behaviors of a group of people can be coherently imagined, together with artifacts and actions.

In a strategic design project, the designer performs a qualitative and interpretative reading of reality, identifying elements that underpin solutions that, in addition to economic value, include aesthetic, cultural and social values. By examining creative ecosystems through the lens of Design—understood as a process of creating devices (artifacts, processes, systems, etc.) to transform reality—we can analyze the interconnected factors that foster productive dynamics among agents, resources, and conditions.

Considering design as a mediator of the planning process within the field of arts is consistent with the expectations and needs of the sector, since design has a careful look at subjectivity and aesthetics, not always present in management and planning worked from other disciplines.

5 Final Considerations

Discussing planning with artists who are already overburdened due to the multiple roles they play in their respective fields may seem like an excessive demand. The current reality is still far from a solution that balances the scarcity of resources for investment by artistic companies and institutions and the lack of time and willingness for strategic and advance planning of their actions.

This analysis aims to point out not only the inclusion of the planning stage within the process of artistic creation and production, but also to consider it in public policies and external investments, so they can guarantee artists not only the viability of their production, but also the maintenance and preservation of their works. This need meets not only the artist's expectations of subsistence, but also the efficiency of the investment made through funding notices, tax incentives and direct financing. In this way, the results can be scaled, reaching a broader audience and generating significant social impact.

Design, as a discipline that articulates both the interests of investors and the needs of the end audience, presents itself as a promising resource for adapting this process and for preserving the artistic and cultural value of productions. Strategic design, although it is not the only possible approach, constitutes an effective tool that can be adapted to the specificities of the field of arts and its subfields, meeting demands that are often not considered in the constitution and structuring of artistic ventures.

This is because there is no single, definitive starting point, but rather the need for a comprehensive analysis of the entire scenario, considering the different agents involved and the desired results so that the objectives are fully achieved. This process is consistent with the very nature of artistic ventures, since the primary interests emerge from the artistic expression of the creator and not from a strict market need.

The question also arises as to how and by which professionally this process of creating and implementing strategic design would be applied in the artistic context. Would it be more viable to include it as a discipline in the academic training of artists in undergraduate or independent courses? Would the constant presence of a strategic designer engaged in the project be essential? Or should this responsibility fall to a director or coordinator within the general planning of the company or institution?

These are fundamental questions that require further study, as this field of study and practice is still in the exploration and development phase. In the future, it is expected that artists themselves, based on this approach, will be able to identify the most viable path for their work. However, given the challenges imposed by the market, it is essential to establish an ongoing dialogue between strategic design and artistic productions, in order to strengthen sustainability and innovation in the cultural sector.

References

1. Chiavenato I (2004) Gestão de pessoas e o novo papel dos recursos humanos nas organizações. Elsevier, Rio de Janeiro
2. Quinn JB (2006) Estratégias para a mudança. In: Mintzberg H, Lampel J, Quinn JB, Ghoshal S (eds) O processo da estratégia: conceitos, contextos e casos selecionados. Bookman, Porto Alegre
3. Porter ME (1986) Estratégia competitiva. Rio de Janeiro, Elsevier
4. Chiavenato I, Sapiro A (2020) Planejamento Estratégico: da intenção aos resultados, 4th edn. Atlas, São Paulo
5. Mendez-Álvarez CE (2019) Elementos para la relación entre cultura organizacional y estrategia. Revista Universidad y Empresa 21(37):136–169
6. Madruga RAFM, Laus-Gomes V (2020) O discurso sobre a estratégia organizacional em uma instituição pública de ensino superior do Distrito Federal. Jornal de Políticas Educacionais 14
7. Brown T (2018) Design Thinking: uma metodologia poderosa para decretar o fim das velhas ideias. Rio de Janeiro, Elsevier, p 239
8. Wu C (2006) Privatização da cultura: a intervenção corporativa nas artes desde os anos 80. Boitempo, São Paulo
9. Benjamin W (2018) A obra de arte na era de sua reprodutibilidade técnica (1936). Trad. Francisco de Ambrosis Pinheiro Machado. L&PM, Porto Alegre
10. Greffe X (2013) Arte e mercado. Iluminuras e Itaú Cultural
11. Thompson D (2012) O tubarão de 12 milhões de dólares: a curiosa economia de arte contemporânea. BEI, São Paulo
12. Guarato R (2018) Dança, dinheiro e mercado: a economia da dança pelo viés historiográfico. In: Guarato R (ed) Historiografia da dança: teorias e métodos. Annablume, São Paulo, pp 39–62

13. Velloso MP (2007) A dança como alma da brasilidade, Nuevo Mundo Mundos Nuevos [En ligne], Colloques, mis en ligne. Disponível em http://journals.openedition.org/nuevomundo/3709 Acessado dia 19 de abril de 2021
14. Barbosa JM (2014) Por uma economia desenvolvimentista da dança. Dissertação de mestrado em Dança. Universidade Federal da Bahia
15. Marion JC (2018) Contabilidade empresarial, 16 edn. Atlas, São Paulo
16. Pacioli L et al (1959) Summa de arithmetica, geometria, proportioni e proportionalitá: Distinctio IX-Tractatus XI (tractatus de computis et scripturis)
17. Iudícibus SD, Martins E (2007) Uma investigação e uma proposição sobre o conceito e o uso do valor justo. Revista Contabilidade Finanças 18:09–18
18. Guedes O (2024) Arte Mercado e Ascensão. Art Economics Fgv Invest nº 41
19. Lazzarato M (2001) Trabalho Imaterial: Formas de vida e produção de subjetividade. Tradução de Mônica Jesus. DP&A, Rio de Janeiro (Original: Lavoro immateriale, 1996)
20. Boltanski L, Chiapello È (2009) O Novo Espírito do Capitalismo. Tradução de Ivone C. Benedetti. Martins Fontes, São Paulo (Original: Le nouvel esprit du capitalisme, 2005)
21. Lupinacci LG, Corrêa JGF (2015) Redes colaborativas de criação em dança: a composição coreográfica na contemporaneidade
22. Luz Júnior M (2019) Balé Jovem de Salvador: uma companhia de formação em dança
23. Martin R (2012) A precarious dance, a derivative sociality. TDR/Drama Rev 56(4):62–77
24. Tenenblat N (2015) Direção e direcionalidade na criação em coletivo. ILINX-Revista do LUME 7
25. Nascimento M, Coelho M, Dellagnelo EHL (2018) Reconhecimento do trabalho artístico na sociedade de consumo. Revista Pensamento Contemporâneo em Administração 12(3):65–78
26. Brant L (2009) O poder da cultura. Editora Peirópolis LTDA
27. Baptista MA (2008) A pesquisa sobre design e cinema: o design de produção. Galáxia. Revista do Programa de Pós-Graduação em Comunicação e Semiótica 15
28. Bomfim GA (1997) Fundamentos de uma teoria transdisciplinar do design: morfologia dos objetos de uso e sistemas de comunicação. Estudos em Design. V. V. n 2, dez
29. Pereira CP (Org.) (2022) Comunicação, Cultura e Tecnologia [recurso eletrônico]. São Paulo: Vecher, 2022. Disponível em: https://doi.org/10.47585/9786584591110
30. Kotler P, Alexander Rath G (1984) Design: a powerful but neglected strategic tool. J Bus Strategy 5(2):16–21
31. Zurlo F (1999) Un modello di lettura per il Design Strategico. La relazione tra Design e strategia nell'impresa contemporanea. Dottorato di Ricerca in Disegno Industriale–XI ciclo. Politecnico di Milano, Milano, p. 247
32. Meroni A (2008) Strategic design: where are we now? Reflection around the foundations of a recent discipline. Strateg Design Res J 1(1):31–38
33. Cain J (1998) Experience based design: toward a science of artful business innovation. Design Manag J 9(4):10–16
34. Geertz C (1978) A interpretação das culturas. Zahar, Rio de Janeiro

Design Systems Applied to Visual Identities

R. Ribeiro, D. Raposo, R. Almendra, and J. Neves

Abstract Today, brands relate to users through an ever-expanding network of online and offline channels, making design systems essential for maintaining visual identity, growth and brand recall. These frameworks combine design rules, reusable modules and established standards, simplifying processes and ensuring the alignment of different teams and devices. This paper traces the development and practical use of Design Systems in shaping Brand Visual Identities, arguing that they evolve into comprehensive Brand Visual Language Systems that unify directives, digital assets and operational workflows. The article combines an organized literature review with a detailed comparative case study of pioneering brand center solutions, namely Frontify, Bynder and Adobe Experience Manager Assets. The results indicate that companies are increasingly combining design systems with digital asset management to master the growing complexities of brand consistency in rapidly evolving global markets. The research highlights effective approaches while identifying existing gaps, demonstrating that flexible Brand Centers, supported by comprehensive Design Systems, improve teamwork, safeguard brand consistency and support rapid adjustments to evolving communication needs. By linking design, engineering and governance, these configurations increase operational efficiency, boost productivity and cultivate lasting innovation in brand management. As such, the findings offer practical guidance for brand leaders, visual designers and academic researchers seeking to strengthen visual identity management amid the dynamics of digital transformation.

Keywords Brand design systems · Brand management · Brand visual identities · Brand centers · Digital asset management

R. Ribeiro (✉) · D. Raposo · R. Almendra · J. Neves
CIAUD, Research Centre for Architecture, Urbanism and Design, Lisbon School of Architecture, Universidade de Lisboa, Lisbon, Portugal
e-mail: r.ribeiro@ipcb.pt

R. Ribeiro · D. Raposo · J. Neves
Polytechnic University of Castelo Branco, Castelo Branco, Portugal

D. Raposo et al. (eds.), *Strategic Design for Sustainable Futures*, Springer Series in Design and Innovation 65, https://doi.org/10.1007/978-3-032-33282-0_14

1 Introduction

A Design System is a set of components, rules, selection and design criteria and style that, when put together, determine appearance, design pattern or language which ensure the production of consistent products.

Through Design Systems you can speed up the design process and allow different people to design consistent and coherent products, facilitating scaling using design patterns and standards that use defined components and can be tested and applied more quickly [1].

As a concept and practice, the Design systems have evolved significantly, keeping pace with constant technological innovation and the growing demand for efficient and scalable solutions in the development of products, services and brands. These systems play a fundamental role in the modern design of brands and the experience of their target audience, providing a comprehensive set of guidelines, principles and resources that ensure consistency, efficiency, scalability, customer recognition, accessibility and innovation across all digital and analog platforms.

From the earliest days of visual communication to today's most sophisticated digital processes, design systems have shaped the way we organize and standardize the visual identity of brands. With historical roots dating back to the era of printing and industrialization, these systems have gained even more relevance with the rise of the digital world in the business context.

Design Systems was inspired by general systems theory, developed by the biologist Ludwig von Bertalanffy and others from the 1930s onwards to explain abstract and physical concepts or realities, with the aim of visualizing and facilitating the understanding and application of knowledge. Although less well known, this concept is also called Atomic Design in association with the composition of matter to form complex organisms (atoms, molecules, organisms, templates and artefacts or pages). In the context of communication design, the concept of System is often used to explain the visual identity of a brand and in User Experience Design it is used to identify, categorize and define design principles that determine the appearance and visual language of a coherent system [2].

Generally speaking, Design Systems is a theoretical and practical concept that results from combining graphic design resources with technological possibilities to respond to brand communication demands.

In the context of Visual Brand Identity, Design Systems seems to be gaining public expression with the transformation of Brand Guidelines, Brand Manuals and Brand Books into Brand Centers, as a response to the digital transformation, adapting to international markets with greater communication needs in terms of quantity of messages and number of media. According to [3] Brand Centers are digital platforms that make it possible to group, organize and manage all the documents, strategy, guidelines, graphic materials, multimedia banks, templates, pre-print files, information and data that are essential for brand marketing and communication and that are needed by internal and external personnel.

The emergence of tools such as Figma and the development of Design Systems 2.0 represent a transition from static approaches to dynamic and customizable methodologies. This evolution makes it possible to create adaptable and consistent experiences in different contexts. As such, design systems not only increase productivity and collaboration, but also play a crucial role in consolidating companies' visual identity and continuous innovation.

In this essay, we will explore the definition of design systems and their importance in maintaining brand consistency, adapting to new digital platforms and increasing brand value, using the explanatory case study methodology referred to by [4], in order to understand how Design Systems, especially Brand Centres, contribute to the consistency and adaptation of the visual identity of brands in the context of digital transformation.

The aim is to analyse the evolution of Design Systems in the context of the visual identity of brands and their adaptation to the digital environment. As well as investigating the effectiveness of Brand Centres as platforms for guaranteeing the visual consistency of brands in different media and their contribution to continuous innovation.

2 A Brief History of Design Systems

Design systems have played an important role in human innovation since the dawn of civilisation. Since the Stone Age, people have applied systematic thinking to their activities, from weaving baskets, producing textiles, hunting tools or even jewellery.

These ancient practices involved repeating patterns, applying symmetry, balance and proportion, even without the use of formal geometry. Our ancestors sought to create consistency, order and beauty in their work, something we still strive for today.

One of the most significant milestones in the modern era was the invention of printing. The invention of movable type is attributed to Gutenberg in 1440, but it was already being developed in Korea, albeit in a more limited and less systematic way. This attribution to Gutenberg, a goldsmith by trade, was due to the improvement of the metal alloys needed for the mass production of precisely cast letters that made movable type viable on a large scale [5, 6].

As [7] points out, with the proliferation of steam presses, there was an exponential increase in the production of printed materials, which facilitated the reproduction of common phrases, images and patterns. Essential elements that profoundly influenced cultural language and stereotypes, contributing to cultural clichés. As certain ideas or expressions began to be conveyed more frequently by the press, they ended up becoming more firmly rooted in the collective consciousness. The rapid dissemination of content gave rise to the concept of "stereotyping", both in the literal sense, through the use of metal "stereotype" plates for rapid reproduction on printing presses, and in the figurative sense known today. This mechanisation of reproduction also contributed to an excessive use of familiar narratives, fixing them firmly in the collective mind. The concept of "cliché" transcended simple block repetitions in print. It has become not only a common vocabulary word, but also a real threat to society.

In 1570, Palladio's 'Four Books of Architecture' presented one of the first cases of systems, including principles and patterns for architectural project [8]. In the same sense, in his book on architecture, "The Timeless Way of Building and A Pattern Language", de 1979, Christopher Alexander presented the concept of pattern design, to refer to a recurring or reusable solution that can be customized and applied to different contexts. In the same sense, a visual identity can be made with pattern designs that can be used in similar, but not identical, situations and contexts, ensuring recognition, familiarity and consistency, but also allowing you to design more products more quickly [8].

Throughout history, we can identify similar patterns of scalability. A classic example is the case of Henry Ford, who although he wasn't the inventor of the automobile, revolutionised it by introducing the assembly line methodology. Ford's innovation was not just limited to the product itself, but to the efficient and scalable system that made the car viable as a product [9].

These examples highlight the importance of innovation and technological development in making products viable and scalable throughout history. But all new technologies have encountered resistance throughout history. The same happened with the first cars in relation to horses.

At the beginning of the twentieth century, the arrival of mechanical assembly lines was criticised by traditional craftsmen in various sectors, who feared the loss of quality in mass-produced products and consequently the loss of their trades.

It was during the first decades of the twentieth century that the first visual identity systems began to emerge, but they were still small, mainly limited to stationary and a few other applications that varied according to the brand's needs. Examples of this are projects by modernist designers such as Wilhelm Deffke, Karl Bùltmann, Hans Karl Michel, Otto Firle e Kurt Schwitters, among others [10].

The terms used in today's digital design, such as "typeface", "font", "interline" and "alignment", have their roots in the printing era. Although they derive from physical printing processes, these terms are still widely used in contemporary design. However, they are often not the most appropriate when it comes to describing digital experiences, such as the use of "page" or "screen". The persistence of print vocabulary in digital design highlights the profound influence that print culture exerts on our understanding of visual communication, even with the transition to digital media [7].

This not only reflects technological advances in communication but also emphasises the historical and cultural importance of the print era within contemporary design. Revealing the complexity of technological evolution and underlining how print culture has shaped our understanding and practice of visual design today.

Between the late 1950s and the 1970s, the expansion of multinational companies and the Olympic Games favoured the development of visual brand identity systems which, also benefiting from modernist thinking and design, became more complex, complete, coherent and consistent. The complexity and scale of visual identity elements, media and communication artefacts reached a huge number, as can be seen in the cases of the Olivetti, Deutsch Lufthansa, Mexico 1968 Olympic Games e Mexico 1968 Olympic Games, Munich 1972 Olympic Games and IBM design [11].

From a technology perspective, in 1968, Douglas McIlroy presented a proposal at a NATO software engineering conference to accelerate the potential of programming and its scalability by making code reusable, which exponentially accelerated innovation in this field [1].

As [7] points out, the 1990s saw a significant leap in digital design with the advance of computers and browsers. Teams began to share reusable standards to create their own design systems, but their adoption was initially slow due to designers concentrating on visual components only.

Despite technological advances, many designers were still stuck in a print-focused mindset. Documents were seen as static pages. Designers used to print out layouts and post them on walls for review. It was only with the widespread adoption of platforms such as Material Design, Storybook and Zeroheight that product teams began to migrate to digital workflows.

In this context, Brad Frost played a key role in documenting and standardising the practice of design systems. Frost [12] proposed a methodology capable of transforming the way we design digital, adaptable and cohesive interfaces.

This approach is based on breaking down interfaces into smaller components (atoms, molecules, organisms, templates and pages), allowing design elements to be better understood and reused, and can be applied systematically to all projects. This standardisation was extremely necessary at the time, when design had little influence in organisations and was often underestimated as a "cosmetic" aspect of product engineering.

The evolution of design tools played a key role in improving efficiency and collaboration in this process.

Over time, design systems have evolved to deal with existing constraints and to overcome the separation between design and development. With the need for scalable design in the digital context, new software and workflows have emerged.

In the last decade, we have seen a significant change with an increase in creativity on the web, driven by social networks and accessible technology.

The impacts of the rise of the Figma application as a design tool have been widely documented. Figma has emerged as a popular choice in interface design due to its comprehensive and integrated approach. By uniting design, prototyping and real-time collaboration capabilities, Figma has solved communication challenges between teams, promoting a more fluid and efficient experience. This transformation not only revolutionised design technology but also set a new industry standard by facilitating effective collaboration between designers, engineers and other stakeholders. Figma quickly became a favourite over other applications and software such as Sketch, InVision and even Adobe XD, thanks to its cloud-native architecture and ease of use [13].

During the COVID-19 pandemic, companies have increasingly adopted digital platforms to facilitate remote collaboration. Tools such as Slack, Zoom and Microsoft Teams gained popularity for holding virtual meetings and teamwork. The use of the cloud as real-time shared storage also became widespread, ensuring that systems could support a global workforce operating from home. Companies realised that they could bring together talent from different locations and work effectively remotely.

But Figma didn't just revolutionise UI design. It also introduced FigJam, a virtual whiteboard that facilitates brainstorming and ideation in real time. This tool is not just restricted to designers, but is accessible to all team members, including marketing professionals and programmers. Figma has become an indispensable tool for entire teams, making it possible to centralise research, knowledge and design in a single place, thus increasing efficiency and the reuse of team resources. In this way, Figma has established itself as a leader in the design market.

In 2021, Figma introduced Variables, which replaced Styles and became a crucial innovation in this regard.

Variables allowed designers to assign tokens to brand attributes, facilitating synchronisation with developer kits, as defined in Material Design. This approach enables Figma to manage dynamic content and themes, such as dark and light modes or multi-brand systems. This possibility of assigning tokens represented a significant leap forward in terms of productivity, allowing designers and developers to control the look of interfaces in real time.

The integration of design and development has been a key objective since the transition to digital media. With a style guide condensed into a single data sheet managed by Figma, changes to the user interface can be made separately from the front-end work.

In addition, the developer kits allow Figma to model objects and components with similar semantic precision to the front-end. This approach further strengthens Figma's role as a reliable source for design.

In parallel with Figma's evolution, Digital Assets Management (DAM) has migrated to the cloud, enabling remote access, real-time collaboration and automatic backups.

Burchill [14] states that this type of system establishes itself as the single, undisputed source of its digital assets, assuming it as the fundamental core of all its visual content. This practice describes the accommodation and ordering of your digital assets in a solution developed specifically to manage these resources. These systems facilitate the management, search and implementation of your digital files in your marketing channels.

One of the most significant benefits of the DAM system is the saving of time and financial resources. It should be considered as a specialised search engine for visual content. Minimising the time spent searching for digital assets frees up space for content creation, campaign planning and the effective fulfilment of other obligations inherent to professional activity.

Another significant advantage of DAM is brand image protection. Many DAM service providers focus on "brand integrity", a crucial element in brand management. The availability of suitable, high-quality images guarantees an optimised experience for customers, thus helping to protect the brand's reputation.

It's common for emerging brands to collaborate with external teams to create creative assets, run campaigns and produce content. With the help of the DAM system, this process becomes more agile and efficient, since all operations can be carried out within the application itself. This feature speeds up the approval process and facilitates collaboration with agencies, freelancers and creators.

Nowadays, companies are facing increasing pressure to innovate quickly in the digital transformation, whether they are large or small [15].

In other words, Digital Asset Management can lead to significant savings in time and money, protect brand image and simplify collaboration with external teams. It is undoubtedly a valuable tool for any organisation that wants to manage its digital assets effectively.

Design Systems with Digital Asset Management integration play a crucial role in this process, enabling integrated and scalable solutions that balance innovation and efficiency. Previously static, these systems have evolved into dynamic structures that link design and development, improving collaboration between teams and providing consistent experiences on a large scale.

This change reflects broader technological and business transformations, influencing designers' practices and organisational approaches to problem-solving. This working procedure is now essential for managing large-scale design in companies. The convergence between design and development facilitated by collaborative tools allows teams to work faster and more efficiently, implementing design decisions with less risk.

3 Importance of Design Systems in the Creation and Management of Visual Identities

The current concept of Design Systems represents the evolution and consolidation of these practices, representing a crucial milestone in the development and implementation of successful products, services and brands. These systems encompass not only tangible elements, such as essential tools and visual standards, but also intangible components, such as brand values and working methodologies.

A Design System goes beyond simple guidelines; it is a complete product that optimises the work of everyone involved in the project. It should be developed iteratively, always with the users' needs in mind. The more it is integrated into the daily lives of everyone involved in the brand perception management process, the more effective it will be. There are tools available to ensure that the system is not just a reference, but a practical, everyday tool that increases productivity and promotes consistency in the final products [16].

As [7] says, as companies evolve digitally, it is becoming increasingly clear that design systems play a crucial role in preserving brand value and ensuring a cohesive digital presence. Their role goes beyond mere efficiency, and they are essential for consolidating and strengthening companies' visual identity in the online environment.

It turned out that Design Systems went beyond a simple design trend. They highlighted a gap in communication between multidisciplinary teams, namely brand managers, designers and developers, which had been observed for a long time and has gained prominence in recent years.

This growing need for standardisation in design arises especially with the transition to digital strategies, making Design Systems even more relevant. They bridge the gap between the printed and digital worlds, establishing a common language that evolves over time. Supported by modern tools, they facilitate the efficient creation and sharing of components. With this progress, companies, especially small and medium-sized enterprises (SMEs), have changed their approach to digital strategies. Today, platforms are not merely support tools, but the core of the product offering, enabling autonomous content management through a content management system (CMS). This transformation has enabled companies to gain control over their digital ecosystems, giving them a competitive edge in the market.

According to [17], Design Systems are a complete source that brings together all the elements necessary for teams to successfully design, implement and develop a product. Both [16, 17] agree that a Design System is an approach that aims to standardise and facilitate the entire creative and productive process in building a product, service or brand.

Exploring Design Systems reveals the importance of defining clear objectives and shared values before implementation. Design principles guide creative decisions, while brand identity encompasses all the visual elements that define it. The components and standards act as fundamental blocks in the design and development process.

In other words, a Design System is a comprehensive source that guides and unifies the creative and production process.

Recent advances in the Figma platform have revolutionised the way design standards are conceived. Now, instead of rigid and static structures, dynamic and highly customisable products are emerging, under the heading of Design Systems 2.0. These systems are not limited to mere design models—they are flexible and based on usability standards. This means that design can evolve continuously, without being trapped in predefined visuals or fixed components.

Gulsen [13] says that Design Systems 2.0 stand out for their ability to be updated, which is one of their greatest assets. When changes are made to a specific element, they can be quickly propagated to all the interfaces in which that element is used, guaranteeing consistency, speed and precision in design updates. This automation not only reduces manual labour but also ensures a more robust user experience across multiple touchpoints.

In addition, the advanced customisation provided by a dynamic design system makes it possible to create personalised experiences for each user. The integration of customer data allows designers to programmatically adjust interface elements, such as colour schemes, content layout or product recommendations, based on individual preferences and behaviour. The standardised code structure makes it easy to integrate these personalised elements into the overall design structure without creating conflicts or inconsistencies.

It is expected that these systems will become more sophisticated, possibly using Artificial Intelligence to automate and improve design processes [18, 19]. This evolution will allow designers and developers to focus on more complex creative challenges, confident that the fundamental elements of their projects are solidly structured.

The flexibility of Design Systems 2.0 also boosts the efficiency of A/B testing. Thanks to the modularity and standardisation of design elements at code level, designers can test multiple variations quickly without the need to create new models or assets. This way, teams can experiment with different design patterns and content strategies, collect performance data and implement changes quickly. As Checker [16] points out, adopting design systems isn't just about incorporating new tools or processes, it's about transitioning to a structured and flexible approach to design. This approach can revolutionise companies' iterative processes, making it possible to quickly optimise interfaces based on user feedback.

If we focus on the challenges facing companies today, we realise that the points of contact have multiplied in the last decade, as have the elements and supports [20–22]. This proliferation demands a robust design system to maintain brand coherence. With this demand, it is increasingly common for there to be internal and external players, making the management process complex, difficult to organise and available to teams at any time. It is essential for these teams to have an integrated, continuous and informed view of all the brand management processes to ensure that the brand experience is consistent, actively contributing to embodying the company's purpose and value proposition.

In other words, the brand is the memory in a consumer's mind. In saying this, we are emphasising the need for the brand to create experiences so that consumers create a brand image in their minds in the right way, whether it's in the experience of consuming the products, the impact of communication, the opinion of a third party on social media or the presence of the products in places of sale [23].

In this respect, brand management is an increasingly complex process, always taking visual perception as its starting point.

On the contemporary stage of visual brand identities, design systems are emerging with increasing vigour, shaping traditional Brand Guidelines, Brand Manuals and Brand Books, which are static documents, into innovative digital BrandCenters that allow dynamic updates, with remote access, customisation of content by market and integrations with other asset management and content creation tools.

These BrandCenters allow companies to control their visual assets, brand guidelines and communication materials in an integrated way, ensuring the coherence of the corporate image in different markets and channels.

As Aggylon [24] says, BrandCenters are like catalysts in the effective management of a brand, performing remarkable functions. One of their main features is the centralisation of resources. A single platform brings together all the brand's assets, including visual identity manuals, digital assets (such as graphic brands, typography, images and videos), communication templates and application guidelines. This grouping facilitates the accessibility and use of these resources by various teams within the organisation, as well as subcontractors, eliminating redundancy of effort.

Using this type of centralised platform ensures robust consistency in all communications and interactions with the public across all channels. This congruence strengthens the brand's identity and ensures that the value proposition is conveyed uniformly, reinforcing a cohesive and consistent image over time.

BrandCenters strive to increase efficiency and productivity by automating workflows and facilitating access to materials. This facility reduces the time spent searching for assets, directing teams to more strategic tasks. In addition, they foster collaboration between different departments by providing universal access to the same guidelines and materials, encouraging joint work and the sharing of best practices.

With centrally managed data and assets, managers get a comprehensive and clear view of the state of the brand. This overview helps them make more informed and strategic decisions. It also promotes the autonomous management of brand components by teams, reducing response times and, consequently, the costs associated with inefficient processes or hiring external resources to create materials.

In other words, BrandCenters are essential for maximising the value of brands and strengthening their presence on the market. They ensure more effective and relevant management of corporate identities.

Many are based on Design Systems, which are both structured and flexible, adapting to different variations quickly without the need to create new models. They are based on performance data, enabling changes to be implemented quickly or even in real time.

As Summa [3] points out, the absence of appropriate tools results in various adversities faced by brand managers, who are forced to invest an excessive amount of time in carrying out mechanical tasks and rectifying errors or inconsistent implementations. This scenario often results in project delays and budget overruns, as well as the implicit danger of potentially damaging the brand's reputation.

4 Cases Studies

Various platforms offer Brand Centre solutions with different characteristics. The digital Brand Center platforms analysed: Frontify, Bynder, Brandfolder, Acquia DAM, Aggylon, Adobe Experience Manager Assets, Canto and MediaValet play a central role in the integrated management of brands' visual identity, particularly in the integration between Brand Center and Digital Asset Management (DAM), responding to the growing complexity of digital ecosystems and the need for consistency in communications. Although each solution has unique features, there is a set of transversal functionalities that define the state of the art in these tools (Table 1).

Common Features

All the tools promote the centralisation of digital assets (graphic marks, typography, colour palettes, templates, promotional materials, among others) in a unified online repository, favouring accessibility and the continuous updating of brand guidelines. This centralisation is supported by robust Digital Asset Management (DAM) systems,

Table 1 Comparative summary of Brand Centre digital platforms

Criterion	Frontify	Bynder	Brandfolder	Acquia DAM	Aggylon	Adobe experience manager assets	Canto	MediaValet
Main Focus	Branding, interactive guidelines	DAM + Brand Guidelines balanced	DAM (rapid asset management)	Technical DAM and asset security	Brand governance and compliance	Asset management integrated with the Digital Experience	Simple and efficient DAM for SMEs	Enterprise DAM with a focus on security and performance
Dynamic Guidelines	Excellent	Good	Limited	Basic	Excellent (modular)	Yes, highly customisable	Limited, document-oriented	Yes, with customised brand portals
Digital Asset Management (DAM)	Good	Very strong	Extremely robust	Robust	–	Yes, high performance DAM	Yes, simple and functional DAM	Yes, high-security enterprise DAM
Template Customisation	Flexible and easy	Available, but with limitations	Limited	Unintuitive	Flexible and automated	Yes, with Dynamic Media	Not native (depends on integrations)	Limited, more focused on asset management
Campaign Management	Limited (basic projects)	Integrated (marketing workflow)	Not a focus	Partial	Strong (campaign control)	Yes, with Adobe Campaign (optional)	Limited	Limited
Governance and Compliance	Medium (control by permissions)	Strong (approval workflows)	Basic	Strong	Very strong (detailed control)	Very high, focused on global compliance	Good, with basic permissions	Very high, with compliance (SOC 2, GDPR)

(continued)

Table 1 (continued)

Criterion	Frontify	Bynder	Brandfolder	Acquia DAM	Aggylon	Adobe experience manager assets	Canto	MediaValet
Collaboration and Feedback	Integrated and fluid	Integrated	Simple	Integrated	Integrated (comments and approvals)	Yes, advanced collaboration	Simple, annotations and markings	Yes, with annotation and asset approval
External Integrations	Many (Adobe, Slack, etc.)	Many (Salesforce, Adobe, etc.)	Basic (Google, Adobe)	Extensive (CMSs, Salesforce)	Customisable (bespoke)	Very broad: Adobe suite, AEM, Salesforce	Basic: Adobe CC, CMSs	Broad: Adobe CC, Microsoft 365, CMSs
Global Scalability	High	High	High	High	High	Very high, Adobe cloud and global CDN	Medium, aimed at SMEs	Very high, Microsoft Azure backbone
Ideal Audience	Creative and agile brands	Medium/large companies	Teams that need DAM agility	Large organisations with strong IT	Companies with a strong marketing structure	Global companies with an advanced digital focus	SMEs and small marketing teams	Large companies and regulated sectors
Companies examples	Nestlé, Budweiser, Grubhub, Tony's Chocolonely, Bosch, Mercedes-Benz, Vodafone,	Five Guys, Alpro, Puma, Spotify, Canon	Russell Stover, Portillo's Bakers Delight, Vermont Cider Co., Lyft, Mastercard, Panasonic	New Balance, Energizer	Heineken, Puratos, Ajinomoto	Philips, Mastercard, Nissan	Harvard University, Ralph Lauren	Experian, Visit Orlando, Hard Rock

which ensure the safe storage, efficient organisation and rapid retrieval of content through search mechanisms structured on metadata, categories and keywords.

Ease of use between the platforms emerges as a common denominator between the solutions, albeit with variations in the degree of complexity, depending on the profile of the users, be they designers, marketing professionals or commercial teams. Frontify [25] stands out for its minimalist interface and low learning curve, favouring rapid adoption even by users with limited technical skills. Similarly, Bynder [26] provides a fluid user experience, albeit with a slight additional complexity due to the range of functionalities it integrates. Brandfolder [27] focuses on operational simplicity, optimising the speed of access and asset sharing, and is particularly suited to marketing teams who value agile solutions. Canto [28] and MediaValet [29] also show a strong commitment to usability, offering intuitive navigation and simplified onboarding processes, albeit with more restricted customisation resources. On the other hand, platforms such as Acquia [30], Aggylon [31] and Adobe Experience Manager Assets [32] are initially more complex, requiring specialised training and are therefore more suitable for organisations with dedicated technical teams.

Interdepartmental collaboration is encouraged across the board by the platforms, either through the possibility of direct comments and reviews on assets, or through the implementation of structured approval flows. At the same time, the management of granular permissions, covering different levels of viewing, editing and downloading, ensures the integrity of information and the protection of digital assets.

In the area of creating communication materials, there is also convergence: all the solutions analysed offer pre-configured templates that democratise content production, allowing non-specialist users to generate materials in line with the brand's visual identity. Frontify leads the way in this regard, offering an advanced Template Library that enables the creation, customisation and dynamic management of templates, without relying on ongoing technical support. This feature reinforces user autonomy and guarantees brand consistency on a large scale. Bynder also has solid template customisation capabilities, integrated with approval flows and brand identity rules, which makes it especially suitable for organisations with decentralised marketing structures. Brandfolder and Canto, on the other hand, offer more basic template functionalities, satisfying standardised needs, but with less dynamism and advanced customisation. Adobe Experience Manager Assets stands out for its ability to create highly personalised templates and for its native integration with the rest of the Adobe suite, although it does require specialised technical skills for its initial configuration. Acquia DAM, Aggylon and MediaValet, on the other hand, offer more limited customisation options, focusing mainly on the management, security and efficient distribution of assets, to the detriment of the dynamic creation of new materials.

Integrations with external systems are another important differentiating factor. All platforms offer connectors to widely used tools such as Adobe Creative Cloud, Salesforce, WordPress and Microsoft 365. However, there is variation in the depth and breadth of these integrations. Bynder, Adobe Experience Manager Assets and MediaValet present particularly robust integration ecosystems, supporting complex workflows and reducing information silos within organisations. Frontify also shows good integration capabilities, especially with creative platforms and content management systems (CMS), although with a greater emphasis on personalisation geared towards brand environments. Brandfolder and Canto offer effective integrations for less complex business scenarios. Acquia DAM stands out for its optimisation for the Drupal ecosystem, while Aggylon favours integrations with document management systems, internal workflows and compliance platforms, which makes it especially attractive for legal and regulatory environments.

Analysing metrics and reports on asset performance (such as downloads, views and engagement levels) is a cross-cutting feature, offering support for evaluating the effectiveness of brand management strategies.

Regarding the pricing model, there is a wide range, directly proportional to the functional depth, customisation capacity and scalability of the solutions. Frontify and Bynder are positioned in the medium to high price range, justified by the integration of comprehensive functionalities such as DAM, dynamic Brand Guidelines and advanced template management. Brandfolder and Canto, on the other hand, offer more affordable models aimed at small and medium-sized companies looking for effective solutions, albeit with less functional complexity. MediaValet has adopted a competitive strategy aimed at organisations that manage large volumes of assets and users. Acquia DAM, Aggylon and Adobe Experience Manager Assets represent larger investments, suitable for multinationals and institutions with strict requirements in terms of compliance, integration and scalability.

It should also be noted that most platforms offer modular and scalable plans, making it possible to adapt the solution to the volume of assets, the number of users and the specific integrations required by each organisation.

Differentiating Features

Despite these common functionalities, each platform is distinguished by specific features that respond to different needs:

Frontify stands out for its strong orientation towards creative collaboration, with direct feedback tools on assets and editing based on flexible templates. Its scalability makes it particularly effective for global brands with decentralised operations.

Bynder is characterised by the breadth of its technological integration, offering deep connections with CRM platforms, project management and content networks. The platform is particularly suitable for rebranding operations, new brand launches and international expansion, reinforcing the simultaneous management of local and global identities.

Brandfolder differentiates itself through its highly intuitive interface and advanced analytical functions, which not only make it possible to manage but also to measure the performance of digital assets. The possibility of creating independent brand portals, adaptable to specific markets, gives it great flexibility for companies expanding geographically.

Acquia DAM integrates digital asset management with product information management (PIM), enabling a broader approach that covers not only visual communication elements, but also technical data associated with products. Its ability to create personalised brand portals suitable for different audiences represents a robust solution for large companies and diversified industrial sectors.

Aggylon, for its part, focuses intensely on the user experience, promoting decentralised management that allows local adaptations without distorting the brand's global identity. It stands out for its dynamic guidelines model (replacing static manuals) and its versatility for multicultural organisations or franchising networks.

In short, the digital Brand Centre platforms studied reflect the evolution of contemporary visual identity management needs, characterised by convergence in essential functionalities, but also by strategic differentiation in terms of scalability, technological integration, analytical management and flexibility to adapt to different markets. The choice of the most suitable platform will therefore depend on the organisation's digital maturity, the complexity of its markets and its strategic priorities in terms of collaboration, product management or personalisation of the brand experience.

5 Conclusions

Undeniably, the last decade has been a period of significant change in the world of design systems, which has had a profound impact on traditional workflows. The emergence of platforms such as Figma, the incorporation of cloud-based structures and the exponential growth of remote working have redesigned the way design teams operate and how digital experiences are created.

However, this evolving scenario has brought with it new challenges for organisations. There are companies that continue to operate with outdated infrastructures that are incompatible with modern design systems. It is imperative to replace these obsolete systems with newer structures in order to stimulate innovation and maintain competitiveness.

The use of Design Systems is crucial in the creation and management of visual identities for products, services and brands. They act as a comprehensive guide that standardises the creative and production process, establishing a bridge between the printed and digital worlds. Design systems play a key role in preserving brand value and ensuring a cohesive digital presence as companies evolve digitally. They also facilitate the efficient creation and sharing of components, thereby boosting productivity.

With design systems gaining prominence in business operations, it is vital to maintain cohesion between teams to ensure the smooth running of dynamic systems. Clear standards and well-defined processes are key to avoiding the fragmentation of design efforts, and it is essential to transform static Brand Manuals into digital and dynamic BrandCenters, with simple principles, up-to-date information and Digital Asset Management.

As design systems evolve, teams need to adapt their skills. Designers will have to become more familiar with data, automation and real-time interfaces, while managers will need to better understand design workflows.

With the advance of design systems, the approach to publishing is also being redefined. Content is becoming more fluid, dynamic and sensitive to user interactions and real-time environments. Organisations must adopt flexible approaches that allow content and design to evolve together.

Despite the growing adoption of the Atomic Design methodology, there are gaps in its practical implementation in organisations, where collaboration between teams can be challenging. The lack of appropriate documentation and training is a major limitation.

It is essential to foster knowledge within organisations to implement Design Systems, along with exploring tools such as Brand Centres that facilitate collaboration and communication between multicultural teams.

The results of these processes have significant implications for design practice, suggesting a paradigmatic shift in the way brand managers, designers, marketers, sales people and developers collaborate and communicate.

The lessons learned from past innovations will be crucial as organisations expand the horizons of the potential of design systems. Those able to embrace more dynamic, integrated and collaborative mindsets will thrive in this ever-evolving landscape. The future of design systems lies in their ability to adapt, personalise and improve user experiences in the new landscape.

Several Brand Centre platforms were analysed and it was found that they all promote the centralisation of digital assets in a unified online repository and have robust DAM systems, with a focus on identity consistency and coherence. However, each platform is characterised by specific features that respond to different needs. The choice of the most suitable platform will depend on the digital maturity of each company, the complexity of its markets and its strategic priorities in terms of collaboration, product management or personalisation of the brand experience.

Acknowledgements The authors would like to acknowledge CIAUD, Lisbon School of Architecture, Universidade de Lisboa, Portugal and FCT—Fundação para a Ciência e a Tecnologia, I.P., under the Strategic Project with the references UIDB/04008/2020 and UIDP/04008/2020.This paper is part of the research for doctorate in design carried out by Rogério Ribeiro.

References

1. Suarez M, Anne J, Sylor-Miller K, Mounter D, Stanfield R (2017) Design systems handbook. Design Better. Co by InVision
2. Raposo D (2025) Design-driven branding: a proposal for a methodological model for strategic and creative brand direction and design of the brand visual identity. In: Perspectives on design and digital communication V. Research, innovations and best practices, 1st edn. Springer, Cham, p 378. https://link.springer.com/book/9783031761553
3. Summa (2019) Brand Center, uma ferramenta para gerir a marca. Brandifiers. https://summa.pt/blog/brand-center-uma-ferramenta-para-gerir-a-marca/
4. Yin RK (1994) Case study research-design and methods. In: Applied social research methods series, vol 5. SAGE Publications, Inc.

5. Füssel S (2020) Gutenberg and the impact of printing, 1st edn. Routledge. https://doi.org/10.4324/9781315253718
6. Reed CA (2007) Gutenberg and modern print culture: the state of the discipline II. Book History 10(1):291–315. https://doi.org/10.1353/bh.2007.a222383
7. Gulsen J (2024a) A brief history of design systems, Part 2. Medium. https://medium.com/@jgulsen_97727/a-brief-history-of-design-systems-part-2-fb00a7af7985
8. Kholmatova A (2017) Design systems: a pratical guide to creating design languages for digital products. Smashing Magazine
9. Tomac N, Radonja R, Bonato J (2019) Analysis of Henry Ford's contribution to production and management. Pomorstvo 33(1):33–45. https://doi.org/10.31217/p.33.1.4
10. Müller J (with Remington, R. R.) (2015) Logo Modernism (J. Wiedemann, ed.). Taschen
11. Raposo D (2008. Design de Identidade e Imagem Corporativa: branding, história da marca, gestão de marca, identidade de marca, identidade visual corporativa. Edições IPCB
12. Frost B (2016) Atomic design. Brad Frost
13. Gulsen J (2024b) A very brief history of design systems, Part 1. Medium. https://medium.com/@jgulsen_97727/a-very-brief-history-of-design-systems-part-1-d661a7c70002
14. Burchill A (2023) What is digital asset management (DAM)? https://www.dash.app/blog/what-is-digital-asset-management
15. Cox B (2024) Your guide to digital asset management for small business. https://www.dash.app/blog/digital-asset-management-for-small-businesses
16. Checker H (2024) Everything you need to know about design systems. Medium. https://waffledesign.medium.com/everything-you-need-to-know-about-design-systems-4273075059ac
17. Ribeiro G (2021) The complete guide about design systems: engineering. CI&T. https://medium.com/ci-t/the-complete-guide-about-design-systems-engineering-cd332cf520f2
18. Curtis N (2017) Practicing design systems. EightShapes. https://medium.com/eightshapes-llc/practicing-design-systems-4029decf9e30
19. McKinsey CM, Singla A, Sukharevsky A (2021) Global survey: the state of AI in 2021. The State of AI in 2021. https://www.mckinsey.com/capabilities/quantumblack/our-insights/global-survey-the-state-of-ai-in-2021
20. Kapferer J-N (2012) The new strategic brand management: advanced insights and strategic thinking. Kogan Page Limited.
21. Kotler P, Keller KL (2006) Marketing management. Pearson Prentice Hall
22. Wheeler A (2018) Designing brand identity: an essential guide for the entire branding team. Wiley
23. Bedendo M (2019) Branding: processos e práticas para a construção de valor. Saraiva Educação
24. Aggylon (with Serrano Giménez, A.) (2023) Las 5 ventajas que te harán dominar el juego de la gestión de marcas: ¿Cómo lo consiguió Cofares? Aggylon Manag Syst. https://aggylon.com/wp-content/uploads/2023/06/ebook-cofares-final-comp.pdf
25. Frontify (n.d.) Frontify: Where brands live—brand management software. Where brands live—brand management software. Accessed 26 Nov 2024, from https://www.frontify.com/en
26. Bynder (n.d.) Bynder—The most intuitive enterprise DAM platform. Bynder. Accessed 26 Nov 2024 from https://www.bynder.com/en/
27. Brandfolder (n.d.) Brandfolder—the most usable digital asset management platform. Brandfolder. Accessed 26 Nov 2024, from https://brandfolder.com/
28. Canto (n.d.) Canto—The leader in digital asset management solutions. Canto. Accessed 27 Nov 2024, from https://www.canto.com/
29. MediaValet (n.d.) MediaValet. MediaValet. Accessed November 27, 2024, from https://www.mediavalet.com/
30. Acquia (n.d.) Acquia DAM | Digital Asset Management (DAM) | Acquia. Accessed 26 Nov 2024, from https://www.acquia.com/products/acquia-dam
31. Aggylon (n.d.) BrandCenter Aggylon | Gestión de marca con Digital Asset Management. BrandCenter Aggylon. Accessed 26 Nov 2024, from https://aggylon.com/
32. Adobe Portugal (n.d.) Adobe Experience Manager Assets—Gerenciamento de ativos digitais. Accessed 27 Nov 2024, from https://business.adobe.com/pt/products/experience-manager/assets.html

The Contribution of Brand Imagery to Sustainable Innovation

Manuela Lorenzon Gastal, Fernando Oliveira, and Daniel Raposo

Abstract This article seeks to inquire about how imagery and sustainable innovation relate to visualizing and communicating sustainable ideas with the potential to influence consumer behavior, namely by catalyzing concepts of identity and the positioning of sustainable brands. This study adopted a non-interventionist methodology, with a component of literature review on key concepts and another of exploratory case study. The data indicates that imagery facilitates the understanding of complex concepts and enables the use of sustainable materials and solutions, which are expressed in a multisensory way. By creating visual representations of ecological brand values, imagery allows for stakeholder engagement with a brand, reinforcing its competitiveness and fostering sustainable innovation. On the other hand, sustainable innovation promotes necessary solutions for addressing economic, environmental, and social challenges, optimizing processes and resource use. Imagery acts as part of a brand interface that fosters the adoption of practices, materials, and conceptually compatible processes, promoting stakeholder engagement, particularly with government bodies, academia, industry, and society, facilitating the development of strategies and the search for responses to regional peculiarities.

Keywords Brand imagery · Sustainable innovation · Brand communication

M. L. Gastal
UNIDCOM/IADE, Unidade de Investigação em Design e Comunicação, Lisbon, Portugal

F. Oliveira (✉)
UNIDCOM/IADE, IADE-Universidade Europeia, Faculdade de design, Tecnologia e Comunicação, Lisbon, Portugal
e-mail: fernando.oliveira@universidadeeuropeia.pt

D. Raposo
CIAUD, Research Centre for Architecture, Urbanism and Design, Lisbon School of Architecture, Universidade de Lisboa & Polytechnic University of Castelo Branco, Castelo Branco, Portugal

D. Raposo et al. (eds.), *Strategic Design for Sustainable Futures*, Springer Series in Design and Innovation 65, https://doi.org/10.1007/978-3-032-33282-0_15

1 Introduction

In an era where sustainability has become a crucial factor in business and consumer decision-making, the role of brand imagery in promoting sustainable innovation is increasingly significant. This article explores the intersection of imagery and sustainable innovation, investigating how visual representations of sustainable ideas can influence consumer behavior and contribute to the positioning of eco-friendly brands. By leveraging imagery, companies are able to communicate complex environmental values in a clear, multisensory manner, effectively bridging the gap between abstract sustainability concepts and consumer engagement. This process not only strengthens brand identity but also encourages a shift towards more conscious consumption patterns, positioning sustainable brands as competitive and desirable in the market.

This study employs a non-interventionist methodology, consisting of a literature review of key concepts and an exploratory case study, to understand how brand imagery fosters sustainable innovation. The findings suggest that imagery serves as a powerful tool to enhance stakeholder engagement by visualizing ecological brand values, creating a deeper connection between brands and their audience. Simultaneously, sustainable innovation drives the development of necessary solutions to address global economic, environmental, and social challenges. The integration of imagery into a brand's interface further supports the adoption of sustainable practices, materials, and processes by engaging key stakeholders such as governments, academia, industry, and society. This multifaceted approach contributes to the creation of tailored strategies to meet regional sustainability goals while reinforcing the brand's market presence.

1.1 The Role of Imagery in Contemporary Society

Santaella and Vieira [31] states that languages permeate the world we live in, and we are immersed in them. In other words, we live in a universe of languages, whether verbal, visual, or of other natures. As human beings in society, we are in constant semiosis, a complex process that occurs through the action of signs.

In this context, where languages permeate every interaction, whether verbal, visual, or otherwise, visual communication emerges as a prominent force in creating contexts where communication between brands and people becomes increasingly direct. Thus, culture comes to define itself more through signs of sensory engagement than through the appeal to the rationalism of traditional representation, which privileges the linearity of writing, as Sodré [33] asserts. The polysemy of images gives them different meanings according to the context in which they are embedded. When we consider the audience's imagination as part of this brand-building system, imagery gains strategic value as an element of brand language.

Focusing on the presence of brands in a consumer society, Olins [27] points out that until around the 1980s, a 'brand' was merely a fast-moving consumer product on supermarket shelves [27]. The need for instant differentiation of one brand from competitors has propelled branding to its contemporary state of the art, where imagery is integral to this context.

When we consider Imagery as part of a brand's Look & Feel, we acknowledge it as the visual language that makes a system proprietary and instantly recognizable, as well as expressing a point of view. This system provides cohesion and differentiation for the brand and comprises elements such as colors, images, typography, and composition—[36].

In addition, Oliveira [28], argues that Imagery is associated with everything related to image/photography and the visual characteristics that belong to a particular Brand, including iconography, illustration, and photographic image.

1.2 The Role of Imagery in Brand Communication

In this context, Imagery has become a substantial part of the brand system, as it contributes not only to the identification of a brand when viewed independently of other symbols but also to its differentiation from competitors. Brands are no longer created solely for identification purposes; the increasing complexity of elements within the system makes the structures dynamic and challenging to control. This flexibility allows components to be added or removed without compromising the brand's identity or public recognition. The more structured the identity system, the easier it is for the audience to identify and connect with the brand.

Thus, a brand identity system is defined by the rules for using its components, which create unique meanings when combined to achieve a cohesive semantic representation. Visual Identity offers coherence and consistency to the brand's products and communication, ensuring its identification and differentiation in the market while also requiring management and systematization through brand guidelines, as Raposo [30] points out.

Imagery acts as a link between corporate rationality and human intuition, as it is embedded with symbols, signs, and meanings that, when aligned with the brand's DNA, create a direct connection to the collective unconscious. Therefore, brand strategy should integrate these elements into Imagery as part of the Brand System.

1.3 Relationship Between Competitiveness, Sustainability, and Visual Communication

In this sense, Imagery has become a substantial part of the Brand System, as it contributes to public recognition and connection with the brand and its differentiation from competitors. As Martins [21] states, a brand is a sensory experience, far more than just a name to be remembered. Similarly, Davis [10] asserts that brands also offer a differentiating factor that facilitates the choice between many competing product variants.

In a competitive landscape among products and services, effective communication and production performance become essential. As Slack [32] argues, competitiveness involves outperforming manufacturing activities relative to competitors. Therefore, if a product is produced better than its competition and meets agreed timelines with customers, organizations gain a competitive advantage.

Production within a sustainable framework is based on competitiveness as well as technical efficiency in production, as Daly [8] suggests. In a sustainable economy, growth is adapted and improved through knowledge, organization, technical efficiency, and wisdom. This process occurs without continuously extracting or adding a greater percentage of material-energy from the ecosystem but rather remains at a scale where the remaining ecosystem can continue to function and renew itself year after year.

Within this context of differentiation, technical efficiency, better performance than competitors, improved production, and meeting agreed deadlines, the consumer emerges as someone increasingly inclined to choose services leaning toward commerce. This choice, according to Sodré [33] is influenced by individual sensitivity. The author (2006), argues that the value of a consumer is assessed by their integration into the corporate-media ethos, where the individual's value lies in the interactive affective exchange with information and communication technologies [33], which serve the goals of organizations.

When a consumer chooses a specific product or service, they are seeking differentiation but are simultaneously influenced by organizations in constant competition for superior manufacturing performance and, consequently, economic gain. As Daly [9] observes, process improvements that contribute to manufacturing efficiency increase the supply of the improved product, leading to price reductions. This reduction opens up new applications for the product as it becomes more affordable, resulting in increased resource consumption for its production, even if done more efficiently. Production scale thus continues to expand, a phenomenon known as the "Jevons Paradox." A "frugality-first" policy tends to generate efficiency as a secondary effect, whereas an "efficiency-first" approach does not promote frugality and, on the contrary, makes it less necessary.

1.4 Relationship Between Innovation, Sustainability, Competitiveness, and the Regional Helix

We understand that competitiveness is intrinsically connected to process optimization and improvements in the production and distribution chain of a given product or service. By adopting an integrated approach, where innovation acts as a driver for sustainability-related actions, organizations can boost competitiveness by improving efficiency, reducing costs, and promoting a socially responsible brand image.

The concept of the Triple Helix focuses on innovation within innovation, exploring how the innovation ecosystem can be enhanced by connecting various stakeholders such as students, researchers, managers, entrepreneurs, and policymakers. This model helps these groups understand the roles of universities, industry, and government in creating and developing "an innovative region" with self-renewal and sustainable innovation capacity, as defined by authors [13].

While each regional development project is "unique, with its own characteristics, certain general elements can be identified, such as the Triple Helix and the entrepreneurial university" [13, p. 14]. In this sense, the Regional Helix concept proposes an approach to sustainable regional development, where collaboration among diverse local actors drives solutions tailored to regional needs and challenges. By fostering collaborative work, sustainable design, and innovation together, this model strengthens regional economies, making them more resilient, competitive, and capable of addressing environmental and social crises while promoting long-term economic growth.

Considering the dynamics proposed by the Regional Helix, the research questions investigated in this article aim to clarify the following:

1. Does brand Imagery contribute to expressing, highlighting, and facilitating the concept and identity of a sustainable and regional brand?
2. Is brand Imagery a catalyst for Sustainable Innovation, particularly through the solutions, materials, and resources adopted?
3. Does brand Imagery enhance stakeholder engagement (consumers, government bodies, academia, industry, and society) with the brand?

Thus, the objectives of this article are:

- To detail the contribution of Imagery in facilitating brand communication comprehension.
- To understand how Imagery can generate sustainable innovation.
- To understand how Imagery fosters stakeholder engagement.

With these aims, we consider the following hypotheses:

We believe that Imagery serves as a powerful tool for increasing stakeholder engagement and amplifying the brand's ecological values, creating a deeper connection with the audience. Moreover, we understand that, through Imagery, it is possible to strengthen the identity of a sustainable brand, enhancing the competitiveness of its products or services and fostering consumer trust, avoiding associations with greenwashing, for example.

The integration of Imagery into the brand interface supports the adoption of sustainable practices, materials, and processes, encouraging collaboration among stakeholders. This multifaceted approach contributes to the development of customized strategies that meet regional sustainability objectives while reinforcing the brand's presence in the market.

2 Literature Review

A comprehensive literature review on Imagery and Innovation in Brand Communication, drawing on authors who discuss Branding, to identify connections between a brand's imagery and its personality in the construction and strengthening of the pillars of sustainability, by integrating economic, human, and environmental values through visual communication. By aligning visual elements with sustainable practices, brands can generate positive impacts that go beyond aesthetics, directly influencing consumer behavior and public perception.

2.1 Imagery and Innovation in Brand Communication

Imagery is an integral part of a brand's Look and Feel, meaning it constitutes the visual language that makes a system proprietary and instantly recognizable, as well as conveying a unique point of view, this system provides cohesion and differentiation to the brand, comprising elements such as colors, images, typography, and composition [35, 36].

Supporting this idea, Oliveira [28] states that Imagery encompasses everything related to image/photography and the visual attributes that are part of a brand's identity, including iconography, illustration, and photographic image. In this sense, Brand Imagery plays a strategic role in differentiating the brand from others and in associating it with sustainable values as part of the brand strategy.

Imagery has great potential to promote a brand's values because, in addition to being directly connected to the brand's DNA, it is linked to the sensory experience. Composed of symbols and signs, Imagery can evoke emotions through its qualities, aligning with aesthetic signs, which ideally follow the pillars of Aesthetic Science but also have the power to materialize and generate interpretive processes, as de Almeida Castro [1] explains.

de Almeida Castro [1] further states that the role of the aesthetic sign is capable of provoking, attracting, or repelling an interpreter through sensory qualities inherent to aesthetic ideals. In this way, Imagery brings visual diversity to the System, encompassing photographic style, illustration, and all that relates to image and visual characteristics that define a particular Brand [19, 28]. Through Imagery, we can connect symbols and signs that convey messages of sustainability, social responsibility, and competitiveness for brands.

2.2 Imagery, Sustainable Innovation, and the Regional Helix

According to Vornholz [34], the concept of sustainability can be traced back to the German forestry industry in the eighteenth and nineteenth centuries. The approach to sustainable development encompasses a wide variety of development objectives, yet it must also remain analytically clear. Vornholz [34], quoting Daly, highlights the distinction between development and growth, where growth refers to the quantitative increase in the economy's scale and physical dimensions, while development involves realizing potential and bringing about structural change. For [34], there are multiple interpretations of sustainable development, some narrower and others broader.

The narrower interpretation focuses solely on the ecological dimension, emphasizing the need for sustainability in managing resources and the environment. It seeks to analyze conditions under which 'optimum' resource management can be achieved [4]. For instance, an ecological objective might be the preservation of biodiversity and ecosystems, while economic objectives might include equitable income distribution and meeting basic needs. This leads to fundamental questions about whether a growth strategy can align with ecological sustainability. Authors with more conservative definitions argue that preserving natural functions is critical, as ecosystems serve as essential input factors for production and consumption. Therefore, a certain level of 'natural capital' must be maintained to ensure availability for future generations.

Conversely, authors who incorporate the economic aspect into development strategies tend to emphasize economic value within their sustainability framework. When considering a broader definition of sustainable development, the concept of the Regional Helix becomes relevant.

The broader interpretation of sustainable development allows a strategic focus on the development of local or regional clusters, which can transform the local economy by gaining a competitive edge. This approach can strengthen the identity of a sustainable brand, enhance the competitiveness of its products or services, and foster consumer confidence. Additionally, it promotes collaboration among stakeholders, driving collective efforts toward sustainable growth and innovation.

The Triple Helix model, which explores the synergistic relationship between universities, industries, and governments, is key to understanding innovation and economic development in today's globalized context [2, 5, 6, 11, 12, 18].

Linton [18] argues that the Triple Helix model has been a pivotal framework for fostering innovation and driving regional development. In this way, the Regional Helix plays a crucial role in promoting sustainable innovation by integrating government, academia, industry, and society in a collaborative effort. Local/regional governments develop public policies and distribute monetary incentives that favor sustainable practices. Academia contributes by producing knowledge and research applied to innovation with a focus on local development, considering surrounding potentials, risks, and opportunities. Industry can adopt and develop sustainable technologies and processes, driving regional economic development while enabling its employees to become advocates for sustainable practices within their communities. Finally, civil society actively participates by demanding responsible policies and solutions and by contributing to the implementation of environmentally responsible practices.

Consequently, the contributions of Imagery to strengthening the pillars of sustainability primarily reside in economic value, as improved Imagery aspects favor more effective communication with stakeholders and the target audience, enhancing transparency of actions, which adds value and strengthens brands.

Additionally, efficient communication has the potential to promote regional projects on a global scale, fostering initiatives and talents beyond major centers and generating financial returns by encouraging sustainable practices.

3 Methodology

This research is based on a qualitative study [22], and follows an exploratory approach [17], supported by a literature review aimed at synthesizing information in order to establish connections between [20]. The goal is to generate a theoretical framework on the areas of visual identity systems, focusing on imagery and the relationship between competitiveness and sustainability in the visual communication of a sustainable brand.

The literature will serve as a foundation for a case study [23, 24], which seeks to verify the possibility of integrating image reading and analysis elements into the branding process in a pragmatic context, where the triangulation of the data obtained is managed during the transition from theory to case studies through the analysis conducted by the researchers [23, 24].

For the case study, online data was collected via the landing page of the NGO project selected for this purpose. According to Creswell [7], in case studies, the research phase is generally based on multiple sources of information, such as observations, interviews, documents, and audiovisual materials [7, p. 73]. In these circumstances, visual analysis was chosen through systematically collected visual materials.

The results presented in this article were obtained through observation, cataloging, systematization, and analysis of the collected material based on a selected significant sample taken from the landing page of the NGO under study, which serves as the object of the case study.

The systematic analysis was conducted using the parameters outlined in the Imagery Analysis model, organized in a checklist format. If the parameters were identified in the NGO's Imagery, they were marked on the table.[1] The table summarizes these parameters in a step-by-step format to streamline the analysis.

The parameters of analysis that compose the model for the Imagery analysis applied to the Great Green Wall NGO case study are based on Martine [16] Synthesis of Meanings Model, which served as a foundation. However, we also systematized analysis categories from several other models, drawing from the works of Frascara [3, 14–16, 25, 26, 29]. By cross-referencing elements from these authors, we identified closely related or similar components within their analysis and categorization systems. We found that in the various analysis and categorization systems studied, the authors sometimes use different terminology to categorize identical or very similar variables.

[1] The table with all the complete parameters is attached to the article.

For the case study, online data collection was conducted through the selected NGOs landing page, following Creswell's methodology (2014). In case studies, the research phase typically involves "multiple sources of information, such as observations, interviews, documents, and audiovisual materials" [7, p. 73]. In this instance, visual analysis was employed through the systematic collection of visual materials collect from the website of the NGO.

4 Case Study

The selection criterion for the NGO in this case study was its notable relevance, as it receives support from various international organizations such as the French Development Agency, the World Bank, the European Commission, the Green Climate Fund, and the African Fund. Additionally, the project directly contributes to 15 of the 17 Sustainable Development Goals adopted by United Nations Member States in 2015.[2]

4.1 The Great Green Wall Project

The selected NGO was The Great Green Wall, the project is a collaborative initiative that evolves 11 Sub-Saharan Countries such as Burkina Faso, Djibouti, Eritrea, Ethiopia, Mali, Mauritania, Niger, Nigeria, Senegal, Sudan and Chad), while 7 other Sub-Saharan African countries are associated with the project (Cameroon, Ghana, Benin, Cape Verde, The Gambia, South Sudan and Somalia). The project is called a Pan African Movement, that started in 2007 with the leadership of the African Union, with the goal to restore 100 million hectares of this degraded land by 2030, sequester 250 million tonnes of carbon and create 10 million rural jobs in the rural areas around the green wall. This Africa's dream of a Great Green Wall dates to the 1970s, when vast swathes of fertile land in a region called the Sahel, which spans the southern edge of the Sahara Desert, started to become severely degraded. This project contributes directly with 15 of the 17 UN Sustainable Development Goals. These 17 Goals adopted by world leaders aim to end poverty, promote prosperity and well-being for all, and protect the planet. The 2030 Agenda was signed by 193 countries part of the United Nations. There are 5 main principles: Planet, People, Prosperity, Peace and Partnership.[3]

The Great Green Wall Frontline project was developed by Pentagram's London Office with Civic and the African Union. The Pentagram Agency was responsible

---

[2] ODS Portugal. (n.d.). *Objetivos de Desenvolvimento Sustentável*. Retrieved October 24, 2024, from https://ods.pt/ods/

[3] ODS Portugal. (n.d.). *The 17 Sustainable Development Goals*. Retrieved October 24, 2024, from https://ods.pt/ods/.

for the development of a new identity system which aims to be easy to recognize and simple to reproduce with limited resources (Figs. 1, 2, 3 and 4).

Fig. 1 A Systematization of the Press Kit Imagery. Retrieve from: https://thegreatgreenwall.org/press-kit#press-kit-internal. Access on 24/10/2024

Fig. 2 Systematization of the History page. Retrieve from: https://thegreatgreenwall.org/history. Access on 24/10/2024

Fig. 3 Systematization of the Home page of the Project. Retrieve from: https://thegreatgreenwall.org/about-great-green-wall. Access on 24/10/2024

4.2 Analyzing Through the Model

We began the analysis by isolating the Imagery from other elements within the brand's visual identity system. After systematically organizing the Imagery extracted from the NGO's website, we initiated the analysis process using the brand DNA diagnostic, gathering information on the Brand DNA to identify relevant brand information such as mission, vision, values, brand essence, positioning, and personas.

In the case of the Green Wall NGO, we were able to connect certain aspects to the Mission and Vision, as being an NGO, the objectives are closely related to the brand's DNA itself. We then moved on to analyze the Context. This parameter allows us to identify what is represented and intended through imagery, as well as to describe the brand's context. In her model, Martine [16] proposes a three-step observation process: description, interpretation, and explanation. Here, description involves identifying the visual elements present in the image, interpretation refers to analyzing the symbolic meaning of these elements, and explanation seeks to understand how the image fits within a broader historical and cultural context. This makes it relevant to analyze the aspects included in the Context parameter.

Next, we analyzed the Colors aspect, identifying primary and secondary colors, with a strong emphasis on green. This directly relates to the project's name and conveys aspects of nature, hence linking to sustainability. In terms of brand DNA, the green tone serves to connect with both the project's purpose and nature itself.

Regarding the parameter of Light, we analyzed the presence of natural and artificial light within the imagery, noting both studio-based indoor images and outdoor

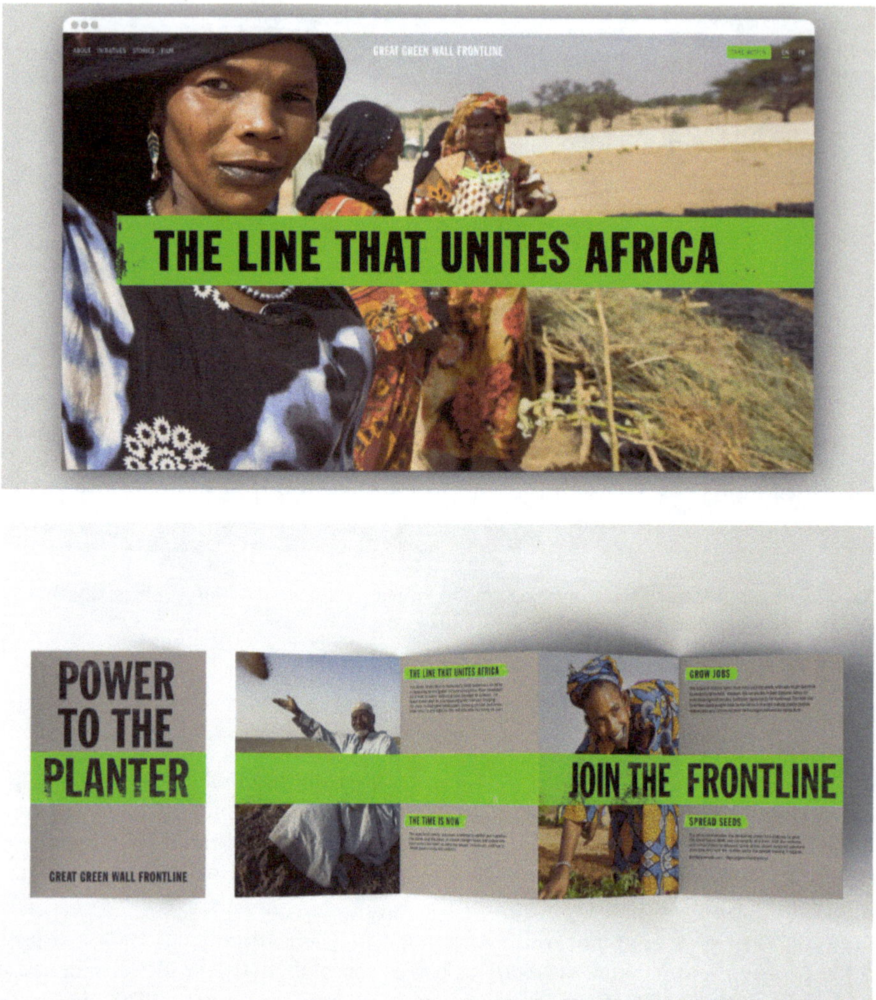

Fig. 4 Imagery developed by Pentagram Retrieve from: https://www.pentagram.com/work/the-great-green-wall-frontline/story. Access on 24/10/2024

images featuring people and the project's location. Natural lighting within the images highlights vibrant colors and contrasts between individuals and the natural background, linking to the Brand DNA/Personality.

In the Objects and Shapes parameter, we noted a variety of objects reflecting the project's context, such as pots, basins with water, carts, adults working, children posing, and solar panels. Among the plastic signs, we observed a contrast between the arid landscape colors and green areas.

Concerning Signs and Styles, we identified the presence of iconic sociocultural messages related to the project environment. Metaphor and synecdoche were noted through rhetorical elements mapped in the project's Imagery.

In the Movement parameter, we highlighted the direction of gaze within the Imagery, showing a dynamic balance of visual elements that convey a sense of spontaneity in the posed photos.

Regarding the Angles used in the project's Imagery, there is a notable variety, depending on the intention to convey specific ideas. Wide shots aim to provide more context, close-ups focus on specific details, and normal angles bring the viewer closer to the people portrayed in the project.

Finally, the Composition analysis aims to synthesize all Imagery elements. We identified other parameters related to the brand's DNA, emphasizing graphic elements that reinforce the projected ideas through imagery anchoring. The presence of horizontal orientation relates to Jungian archetypes, identified in the Imagery analysis, representing the caregiver through the depiction of children and elders, as well as symbolizing a concern for the ecosystem and nature. There is no predominant framing style in the Imagery, which employs various angles and perspectives as needed.

5 Conclusions

In conclusion, Imagery plays a crucial role in sustainable innovation by facilitating the understanding of complex concepts and promoting engagement with ecological values through visual communication. By creating visual representations that reinforce the identity of sustainable brands, Imagery contributes to market competitiveness and encourages more conscious consumption practices. The ability to integrate visual elements with sustainability principles not only strengthens the brand's message but also creates an emotional connection with consumers, increasing their willingness to adopt behaviors aligned with sustainable development. Imagery can be considered a catalyst for Sustainable Innovation, as it incorporates visual elements that reinforce sustainable practices, providing clear communication of the brand's environmental commitment. This approach not only enhances positive environmental impact but also drives the development of sustainable processes and practices, contributing to the continuous innovation and strengthening of brands' ecological image.

Thus, we conclude that Imagery is a powerful tool for engaging stakeholders and reinforcing ecological values, as well as providing efficient visual communication that supports the adoption of sustainable practices. This multifaceted approach allows for the development of strategies aligned with regional sustainability goals and strengthens the brand's presence and competitiveness in the market.

Looking ahead, it is expected that visual communication will continue to gain relevance in promoting sustainability, with brands increasingly using imagery to effectively convey their ecological values. Best practices in communication within the context of sustainability will adopt a more integrated approach, where visual aesthetics and environmental commitment go hand in hand. Innovation in the use of sustainable materials and in the creation of authentic visual narratives will be key factors in addressing global challenges, while also providing new opportunities to develop strategies that create a positive impact for both society and the environment.

Acknowledgements This study stems from the thesis Design Principles Applied to Semiotics in Brand Imagery: The Cases of Repertory Dance Companies.

References

1. de Almeida Castro MC (2009) Web design: Guia de Produção e análise [Tese de Doutorado apresentada no Programa de Estudos Pós-Graduados em Comunicação e Semiótica, Pontifícia Universidade Católica de São Paulo]. https://repositorio.pucsp.br/jspui/handle/handle/5265
2. Amaral M, Cai Y (2023) A decade of triple helix journal—achievements and challenges. Triple Helix 9(3):239–243. https://doi.org/10.1163/21971927-12340008
3. Arnheim R (2000) Arte e Percepção Visual—Uma psicologia da visão criadora (I. T. de Faria, Trans.; 13ª). Editora Pioneira
4. Barbier EB, Markandya A (2017) the conditions for achieving environmentally sustainable development. In: Mollica D, Campbell T (eds) Sustainability. Routledge, pp 659–669. https://doi.org/10.4324/9781315241951
5. Cai, Y., & Etzkowitz, H. (2020). Theorizing the Triple Helix model: Past, present, and future. *Triple Helix Journal*, 1–38. https://doi.org/10.1163/21971927-bja10003
6. Carayannis EG, Campbell DFJ (2009) 'Mode 3' and 'Quadruple Helix': toward a 21st century fractal innovation ecosystem. Int J Technol Manag 46(3/4):201. https://doi.org/10.1504/IJTM.2009.023374
7. Creswell JW (2014) Research design: qualitative, quantitative, and mixed methods approaches, 4th edn. SAGE Publications
8. Daly HE (2004) Crescimento sustentável? Não, obrigado (V. R. Alves, Trans.). Ambiente Sociedade 7(2):197–202. https://doi.org/10.1590/S1414-753X2004000200012
9. Daly HE (2006) Sustainable development—definitions, principles, policies. In: Keiner M (ed) The future of sustainability. Springer Netherlands, pp 39–53. https://doi.org/10.1007/1-4020-4908-0_2
10. Davis M (2006) More than a name: an introduction to branding. AVA Academia, Distributed in the USA and Canada by Ingram Publisher Services. http://public.ebookcentral.proquest.com/choice/publicfullrecord.aspx?p=4654144
11. Etzkowitz H, Brisolla SN (1999) Failure and success: the fate of industrial policy in Latin America and South East Asia. Res Policy 28(4):337–350. https://doi.org/10.1016/S0048-7333(98)00077-8
12. Etzkowitz H, Leydesdorff L (2000) The dynamics of innovation: from National Systems and "Mode 2" to a Triple Helix of university–industry–government relations. Res Policy 29(2):109–123. https://doi.org/10.1016/S0048-7333(99)00055-4
13. Etzkowitz H, Zhou C (2018) The triple helix: University-industry-government innovation and entrepreneurship, 2nd edn. Routledge
14. Frascara J, Meurer B, van Toorn J, Winkler D (1997) User-centred graphic design: mass communications and social change. Taylor & Francis
15. Gervereau L (2007) Ver, compreender e analisar as imagens. Edições 70 Ltda
16. Joly M (2012) Introdução à análise da imagem. Edições
17. Kumar V (2013) 101 design methods: a structured approach for driving innovation in your organization. Wiley

18. Linton G (2024) Triple helix dynamics and hybrid organizations: an analysis of value creation processes. J Knowl Econ. https://doi.org/10.1007/s13132-024-01911-2
19. Lupton E (2011) Graphic design thinking: beyond brainstorming, 1st edn. Princeton Architectural Press, Maryland Institute College of Art
20. Martin B, Hanington BM (2012) Universal methods of design: 100 ways to research complex problems, develop innovative ideas, and design effective solutions. Rockport Publishers. https://edisciplinas.usp.br/pluginfile.php/5646766/mod_resource/content/1/MARTINHANINGTON_Universal-Methods-of-Design.pdf
21. Martins J (1999) A natureza emocional da marca: como encontrar a imagem que fortalece sua marca. Negócio Ed
22. Mason J (2017) Qualitative researching, 3rd edn. SAGE Publications
23. Muratovski G (2016) Research for designers: a guide to methods and practice. Sage Publications Ltda
24. Muratovski G, Friedman KS, Norman DA, Heller S (2022) Research for designers: a guide to methods and practice, 2nd edn. SAGE Publications
25. Niemeyer L (2007) Elementos de semiótica aplicados ao design. 2AB
26. Noble I, Bestley R (2016) Visual research: An introduction to research methods in graphic design, 3rd edn. Fairchild Books, an imprint of Bloomsbury Publishing Plc
27. Olins W (2008) The brand handbook. Thames & Hudson
28. Oliveira FJMS (2015) Diagramas and Marcas: Contributos sobre a Utilização de Diagramas na Conceção e Análise do Discurso Visual das Marcas [Doutoramento em Design, Universidade de Lisboa, Faculdade de Arquitetura]. http://hdl.handle.net/10400.5/13974
29. Poulin R (2018) The language of graphic design revised and updated: an illustrated handbook for understanding fundamental design principles, 2nd edn. Rockport Publishers
30. Raposo D (2012) La letra como signo de identidad visual corporativa: Codificación y descodificación visual del sistema de identidad [Doutoramento em Design, Universidade de Lisboa, Faculdade de Arquitetura]. http://hdl.handle.net/10400.11/1424
31. Santaella L, Vieira JA (2008) Metafísica Como Guia Da Pesquisa: Uma Proposta Semiótica E Sistêmica. Mérito Editora
32. Slack N (2002) Vantagem Competitiva Em Manufatura. Atlas
33. Sodré M (2006) As Estratégias Sensíveis—Afeto, Mídia E Política. Editora Vozes
34. Vornholz G (1994) The sustainable development approach. Intereconomics 29(4):194–198. https://doi.org/10.1007/BF02926438
35. Wheeler A (2018) Designing brand identity: an essential guide for the entire branding team, 5th edn. Wiley
36. Wheeler A (2019) Design de Identidade da Marca: Guia essencial para toda a equipe de gestão de marcas (F. A. Da Costa, Trans.; 5ª ed.). Bookman

Instagram: Private Universities and Content Strategies

Maria Luisa Costa, Mariana Simões de Castro, and Ana Sofia Olivença

Abstract This study seeks to analyze the use of Instagram by a selection of private universities in Portugal to understand which content strategies drive user interaction and which hinder it. Through qualitative and quantitative analysis, this research examined key metrics such as post frequency, content types, and visual elements across eight private universities over two months (May–June 2024). The data were collected from public Instagram profiles, excluding Stories due to privacy restrictions. The study reveals that authentic and high-quality posts, especially those that humanize the institution, generate a sense of belonging and promote greater community engagement. Universities that balance promotional content with community-focused narratives receive higher levels of interaction, while those relying heavily on promotional material see a decrease in user engagement. This paper discusses the key findings on the effectiveness of communication strategies and offers insights into optimizing the social media presence of academic institutions. The findings suggest that authenticity, quality, and community-centered narratives are key drivers of engagement, while excessive promotional content may reduce user interaction.

Keywords Instagram strategy · Visual communication in social media · Private universities · User interaction metrics

1 Introduction

In today's digital age, social media platforms, particularly Instagram, play a crucial role in institutional communication and brand building. For private universities, which often rely on student enrollments for financial sustainability, Instagram provides a valuable platform to showcase their academic offerings, values, and

M. Luisa Costa (✉)
Centro de Investigação em Arquitetura Urbanismo e Design (CIAUD), Lisboa, Portugal
e-mail: design.luisacosta@gmail.com

M. Simões de Castro · A. Sofia Olivença
Instituto Superior Miguel Torga, Coimbra, Portugal

D. Raposo et al. (eds.), *Strategic Design for Sustainable Futures*, Springer Series in Design and Innovation 65, https://doi.org/10.1007/978-3-032-33282-0_16

success stories in a visually appealing way. As the preferred social network for the 18–24 age group, Instagram has become an essential tool for universities aiming to engage and maintain an active online community.

Private institutions, facing the ongoing challenge of securing enrollments, adopt various strategies to promote their programs and strengthen their academic reputation. These strategies often include the promotion of academic achievements, public events, and personal success stories of students and professors. The goal is to create a prestigious image that resonates with prospective students while leveraging the visual nature of Instagram to attract engagement.

The motivation for this study arises from shortcomings observed in the social media strategy of the Instituto Superior Miguel Torga (ISMT), particularly in its Instagram presence. ISMT's reliance on overly promotional content with limited engagement led to a broader investigation into how other private universities use Instagram to engage their audiences. This study compares content strategies and user interactions across eight private universities, providing a comprehensive analysis of effective communication practices in the academic sector. The study analyzes and compares how a sample of private universities in Portugal use Instagram, employing a mixed-methods approach with quantitative and qualitative data collected from posts made over a two-month period on this social network.

2 Contextualization

Private universities in Portugal face significant challenges in attracting students, as they heavily depend on enrollment numbers for their financial sustainability. As a result, these institutions have adopted increasingly robust and aggressive communication and marketing strategies. One of the main channels used to reach potential students is Instagram, a social media platform that has stood out for its ability to capture the attention of young audiences.

According to the Direção-Geral do Ensino Superior [1] there are currently 93 private higher education institutions in Portugal, of which 36 belong to the university sector and 57 to the polytechnic sector. Most of these institutions use Instagram to promote their academic activities, events, and success stories to attract new students and strengthen their institutional reputation.

In this context, Instagram has become a fundamental platform for private universities, offering an effective way to communicate with their target audiences, especially the young demographic aged 18–24, which constitutes the majority of users on this social network [2]. Therefore, universities need to understand which communication strategies are most effective on Instagram, particularly regarding the frequency and type of content published, in order to maximize the impact of their digital marketing efforts.

Due to its visual nature and young target audience, private universities use social networks intensively, and Instagram has become a fundamental tool for their self-promotion.

Thus, this study seeks to analyze the strategies adopted by various private universities on Instagram, evaluating the types of posts and their impact on the audience. An additional point that was not addressed in this study but **holds significant potential** for future research is the comparability of the hashtags used by universities to target potential students and increase institutional visibility.

3 Theoretical Framework

According to Kemp [3], data published by the leading social media platforms at the beginning of 2024 indicate that Portugal has 6.64 million social media users aged 18 or over, representing 77.1% of the adult population in the country. On the other hand, the OberCom report [2] states that Instagram is the preferred social network for the 18–24 age group, both for general use and for news consumption, showing significant growth. These findings confirm the earlier observations in the OberCom report [4], which identified Instagram as the fastest-growing social network, with a 41.6% increase in users between 2015 and 2023. In the 18–24 age group, this growth is even more pronounced, with 72% of young people using Instagram in 2023, representing an increase of 42.7% compared to 2015.

The visual nature of Instagram, designed for photo sharing [5], coupled with the speed at which information is disseminated, makes the content published on this platform particularly relevant for studies in the design field.

Instagram's success is intrinsically linked to its ability to evoke visual nostalgia inspired by the aesthetics of old Polaroids [5]. Chester [6] goes further, stating that Instagram adopts the stylistic language of Polaroid to suggest instantaneity, sociability, and nostalgia through instant capture and sharing of ephemeral moments. In other words, Instagram assumes the social role in the digital reproduction era that Polaroid held decades before: producing instant photos meant to be shared with friends and family. Chester [6], citing Susan Buck-Morss (1991) and her work on Walter Benjamin, highlights that new technologies precisely imitate the old forms they replace, emphasizing that both Instagram and Polaroid are social models of photography designed for the enjoyment of the present moment.

Thus, due to its characteristics and popularity among younger age groups, Instagram offers universities the opportunity to build online communities and self-promote at reduced costs. For this reason, universities have generally adopted Instagram, a practice that has led to several studies analyzing the posts and strategies of universities on Instagram. The study by Moreno-Alabarracín & Blanco-Sánchez [7] and Alcobea et al. [8] demonstrate that private universities predominantly use Instagram as an institutional communication channel, primarily publishing informational and promotional content, with a lack of specialized content adapted to the visual and dynamic characteristics of this social network.

The study on Spanish universities [8] mentions that universities do not allocate the necessary resources for the proper management and use of the platform, underestimating the importance of a more elaborate strategy and the creation of specific content for Instagram, noting that profiles do not maximize the platform's potential to generate interactions and strengthen institutional visibility.

The study that analyzed how Instagram is used to build communities by the five best universities in the world (Harvard, Stanford, Cambridge, the Massachusetts Institute of Technology (MIT), and the University of California, Berkeley) [7], concludes that there is no direct relationship between post frequency and interactions. Harvard, the university that publishes the most, receives the fewest interactions, while the University of California Berkeley, which publishes the least, achieves the best results in terms of interactions. In this sense, they note that content quality outweighs quantity, with institutions that publish less but offer appropriate and aesthetically pleasing content achieving greater user engagement while enhancing the institution's prestige.

Humanizing posts is one of the most effective strategies for universities on social media [7]. Posts featuring personal stories, events involving students and faculty, or the institution's cultural initiatives tend to generate more interactions.

This humanization allows universities to craft a more engaging narrative and an emotional connection with their followers, fostering a sense of belonging. In the university context, capturing and sharing real moments, especially during academic events and in students' daily lives, has proven to be one of the most effective strategies for generating engagement.

4 Methodology

Previous studies in the marketing area and institutional communication on social media provided us with a theoretical foundation and a conceptual model that helped define the content categories and analysis parameters, which we adapted to the reality observed in Portuguese universities.

Thus, this study adopts a mixed-method approach, integrating quantitative and qualitative methods. Data was collected by directly observing posts from eight private universities on Instagram from May to June 2024. The publicly available posts on Instagram were analyzed. Stories were excluded due to privacy restrictions imposed by the platform; therefore, Stories and their respective reactions are not included in this study.

The data were manually collected from Instagram profiles, categorized in a spreadsheet, and supplemented with a data visualization program.

4.1 Selection Criteria for Universities

Given the high number of private institutions in Portugal, we established criteria that allowed us to collect data, evaluate the information, and systematize it, transforming it into intelligible and comparable material.

We included in our sample universities offering at least ten courses in diverse fields and with at least one course classified under code 213—Audiovisuals and Media Production, by CNAEF, according to Portaria n. 256/2005, of March 16th [9], such as Communication Design and Multimedia courses. This last selection criterion is related to our field of study, as we consider that these universities bear additional responsibility in terms of external communication.

As a result, the sample of this study comprises the following higher education institutions:

- Instituto Superior Manuel Teixeira Gomes (ISMTG)
- Instituto Superior Miguel Torga (ISMT)
- Universidade Europeia (UE)
- Universidade Lusófona Porto (ULP)
- Universidade Lusófona Lisboa (ULL)
- Universidade Portucalense (UP)
- Universidade da Maia (UM)
- Universidade Católica (UC).

4.2 Time Frame

The analysis period has covered the last two months of the academic year (May and June 2024), coinciding with the end of the academic year and the preparation for the next one. These months were chosen due to their high potential for Instagram activity, including posts related to the end of the current academic year, preparation for the next year with the respective promotion of institutions, and academic events such as parties and graduation ceremonies.

4.3 Characterization of Universities

Once the time frame was established and the universities selected, we characterized them by identifying their Instagram pages, page creation date, number of followers, and posting frequency during the analysis period, as shown in Table 1.

From the summary analysis of this table, it is observed that the Instituto Superior Miguel Torga (ISMT) is the oldest institution (1939) and joined Instagram later (2020), having the lowest number of followers (1269) and the highest number of posts (187). The Instituto Superior Manuel Teixeira Gomes (ISMTG) is the institution closest to ISMT in terms of followers (1569), but it has the lowest number of posts (4). However, the Universidade Europeia (UE) stands out as the institution with the highest number of followers (20,600), making 57 posts during the same period. Following UE is the Universidade Católica (UC), the second oldest private university in our sample (1967), with 10,900 followers and 31 posts during the study period.

Table 1 Analysis of the institutions

Institution name	Instagram name	Foundation	Instagram page creation date	Posts number (May 1–Jun 30)	Followers number (Jul, 17–2024)
ISMTG	ismat_institutosuperior	2004	Jan/2014	4	1569
ISMT	ismt.coimbra	1936	Feb/2020	187	1269
UE	ueuropeia	2013	Sep/2014	57	20,600
ULP	ulporto	2005	Apr/2016	8	8290
UP	universidadeportucalense	1986	May/2015	63	9448
UM	umaiaoficial	1991	Apr/2016	15	8191
UC	universidadecatolicaportuguesa	1967	Nov/2016	31	10,900
ULL	Ulusofona	1998	Mar/2013	28	16,500

4.4 Structure and Analysis Categories

To assess the communication of the universities, we monitored and analyzed the total number of posts made by each institution and categorized them according to the types of content observed, dividing them into seven major categories:

- **Institutional**: Posts related to essential academic activities and promoting the involvement and participation of students, faculty, and the general public in the institution's initiatives (exam information, job opportunities, academic life, thesis defenses, exhibitions, workshops, open classes, etc.).
- **Promotional**: Marketing and advertising actions aimed at attracting new students and promoting the institution's courses and programs (promotion of courses, special access contests, facilities, etc.).
- **Recognition:** Posts highlighting the achievements and successes of students, faculty, or alumni (awards received, participation in conferences, research publications, success stories, etc.).
- **Information/Activities**: News or topics of interest to the community (curiosities, generalities, commemorative dates, etc.).
- **Internationalization**: Initiatives that promote international cooperation and academic mobility (exchanges, Erasmus programs, etc.).
- **Participation in Events**: More subtle marketing strategies to enhance the institution's image and reputation (participation in industry events, collaborations with companies and organizations, study visits, etc.).
- **Social Responsibility**: Initiatives by the institution that demonstrate social, environmental, or political commitment (volunteer projects, sustainability campaigns, social inclusion programs, etc.).

Based on this definition, it was possible to evaluate the level of interaction generated by each content category, identifying which were most effective in follower engagement. The categories were developed based on the principles of Katz, Blumler, and Gurevitch [10] who argue that media fulfill essential functions, responding to users' motivations for seeking information, entertainment, and social interaction. In addition, Krippendorff's [11] Content Analysis Methodology was applied, which systematizes large volumes of qualitative data. The process included creating a coding system to group posts by categories and qualitative and quantitative analysis to identify visual patterns and measure the impact of the interactions generated by each category.

From the analysis of the particularities of the posts, it was also possible to define some typologies based on visual characteristics, to conduct an analytical systematization, relating the typologies to the impacts generated on followers, which were measured by the number of interactions.

The typologies defined by visual characteristics were:

- **Organization**: Refers to the number of images or videos per post, i.e., Unique posts or Carousel posts.
- **Source**: Refers to the type of visual content classified as Original, content created by the institution precisely to convey a specific message; Stock Images: posts containing visual content from stock image repositories; Mixed, posts that combine original elements and stock images; Other: use of long texts and tables from written communications that do not fit the previous sub-typologies nor the predominantly visual dynamics of Instagram communication.
- **Media**: Refers to the medium through which the content was produced and presented, classified as Photography: static image posts; Video: video content; Text: posts focused on textual content; Combined: posts combining text, images, and/or videos.

5 Results

Total Posts by Category

The staked horizontal bar chart (Fig. 1) provides a detailed analysis of the 393 posts made during this period, revealing significant variation in the activity of the universities. Instituto Superior Miguel Torga (ISMT) was the most active institution, with 187 posts (an average of 3 per day). Universidade Portucalense was the second most active, with 63 posts (an average of 1 per day), followed by Universidade Europeia with 57 posts. On the other hand, Instituto Superior Manuel Teixeira Gomes posted only four times during the same period (about one post every 15 days), which, as we will see in Fig. 2, does not necessarily reflect the number of interactions obtained.

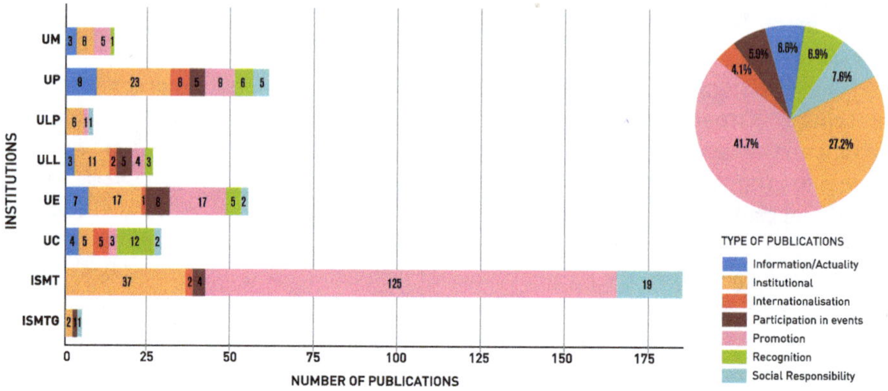

Fig. 1 Stacked horizontal bar chart—Relationship between the total number of posts per institution and the type of categories, showing the distribution of posts in each category. Pie chart—Percentage distribution of total posts by category, highlighting the proportion of each type of content in relation to the total

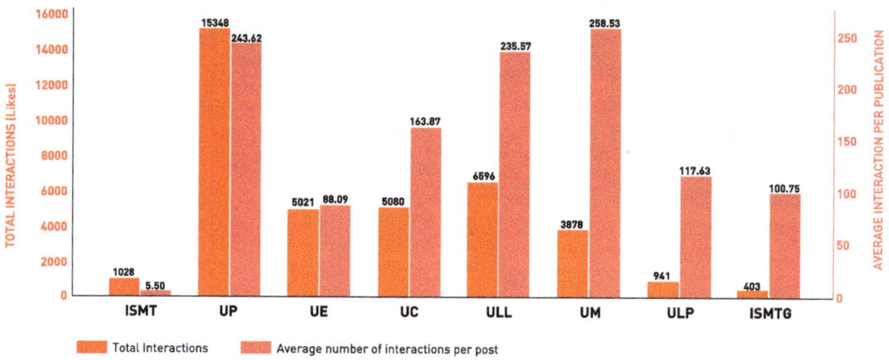

Fig. 2 Clustered bar chart—Comparison between the total and average interactions per post for each institution

Based on the data presented in the pie graph (Fig. 1), most posts are promotional, representing 41.7% of the total. Institutional posts rank second, at 27.2%, while internationalization and participation in events are less frequent, at 4.1 and 5.9%, respectively. Despite the predominance of promotional content, universities such as Universidade Portucalense, Universidade Europeia, and Universidade Católica present a more balanced distribution across different content categories, demonstrating a more varied communication approach.

Interactions per Post

As we can see in the clustered bar chart (Fig. 2), the impact of the posts was measured by the interactions generated (likes and comments), showing considerable variation between the institutions in the average number of interactions per post. Although the most active institution, Instituto Superior Miguel Torga (ISMT), accumulated only 1,028 interactions across 187 posts, resulting in an average of 5.5 interactions per post. In contrast, Universidade Portucalense, with 63 posts, obtained 15,348 interactions, corresponding to an average of 243.6 interactions per post. With just four posts, Instituto Superior Manuel Teixeira Gomes (ISMTG) garnered 403 interactions, resulting in an average of 100.75 interactions per post.

Relationship between Content and Interaction

The clustered bar chart (Fig. 3) relates the different types of content with the interactions, showing that promotional posts, although the most frequent (164), generated a total of 7,072 interactions, resulting in an average of 43.12 per post, the lowest among the analyzed categories. In contrast, institutional content posts (107) obtained 14,282 interactions, with an average of 133.48 interactions per post, indicating a more significant impact.

It is also worth highlighting informational/activities content, which, despite being less frequent (26 posts), proved highly effective, generating 4,958 interactions, with an average of 190.69 interactions per post, the highest average among all categories.

Other types of content, such as recognition (27 posts) and participation in events (23 posts), also proved effective in engaging users, with averages of 173.81 and 151.65 interactions per post, respectively. Posts related to internationalization (16) and social responsibility (30) generated fewer interactions, with averages of 134.88 and 54.80 per post, respectively.

These data show that, although promotional posts are the most frequent, they have the lowest average interactions per post. On the other hand, content related to information/activities and recognition generates significantly higher engagement per post, suggesting that content offering more informational value or social recognition attracts more user interactions.

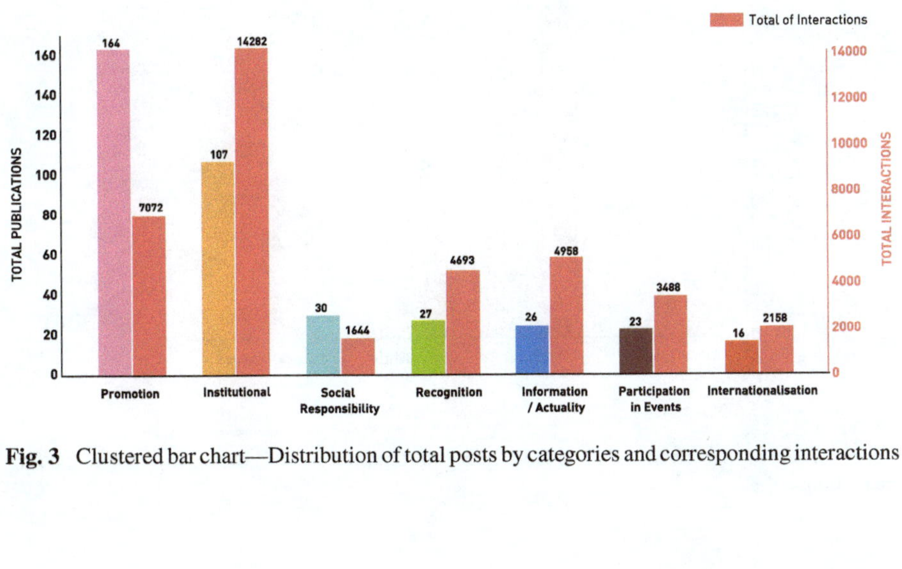

Fig. 3 Clustered bar chart—Distribution of total posts by categories and corresponding interactions

Visual Characteristics and Performance

The visual characteristics of posts are crucial to their success but often difficult to evaluate quantitatively. To overcome this challenge, we analyzed the interactions generated by Media (Fig. 4), Source (Fig. 5), and Organization (Fig. 6).

Interactions by Media

The clustered bar chart (Fig. 4) presents the relationship between the media used in the posts—photography, video, text, and combined formats—and the interactions generated. Photography was the most effective medium, generating 26,126 interactions across 127 posts, with an average of 205.71 interactions per post. Although

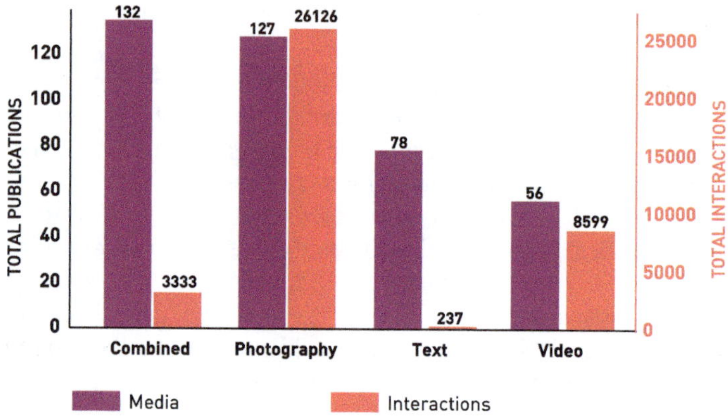

Fig. 4 Clustered bar chart—Comparison between the total posts by media and corresponding interactions

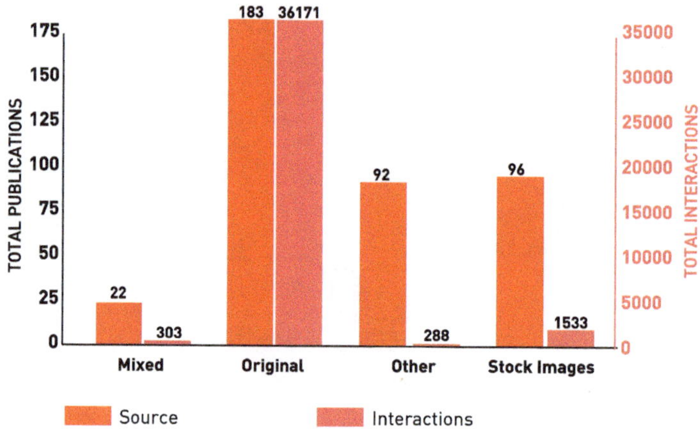

Fig. 5 Clustered bar chart—Comparison between total posts by source and corresponding interactions

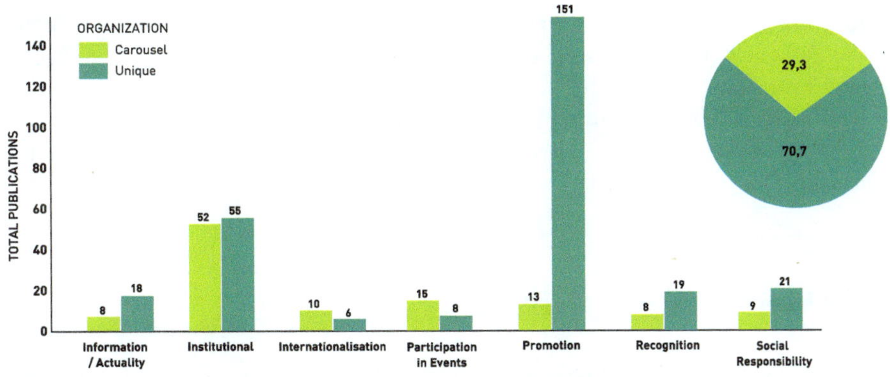

Fig. 6 Clustered bar chart—Distribution of total posts by content category and type of organization. Pie chart—Percentage distribution of the type of organization, highlighting the proportion of each type in relation to the total

used less frequently (56 posts), the video also demonstrated high potential, with 8,599 interactions and an average of 153.55 interactions per post. Combined formats (using more than one medium) were the most utilized, with 132 posts, but generated only 3,333 interactions, resulting in an average of just 25.25 interactions per post. Finally, text was the least effective, with 237 interactions across 78 posts, corresponding to an average of 3.04 interactions per post.

Interactions by Source

As for Source, clustered bar chart (Fig. 5) shows that original content, defined as any image or photograph specifically created to convey a specific message or depict authentic events from the institution, generated the best results, with 36,171 interactions across 183 posts, resulting in an average of approximately 197.68 interactions per post.

Posts that used stock images, although common (96 posts), garnered only 1,533 interactions, an average of 15.97 interactions per post. Content labeled as Other (which does not fit the previous sub-typologies or the predominantly visual communication dynamic of Instagram) accumulated 288 interactions across 92 posts, resulting in an average of 3.13 interactions per post. Lastly, Mixed content (posts that combine different types of content) registered 303 interactions across 22 posts, corresponding to an average of 13.77 interactions per post.

Organization by Category

Clustered bar chart (Fig. 6) demonstrates a significant prevalence of unique images in promotional posts, with 151 (92.6%) using unique images and only 13 (7.4%) using carousels. Despite having far fewer posts, unique images remained the preferred option in the information/activities (88.5%) and recognition (81.5%) categories.

A balanced distribution was observed regarding institutional posts, with 52 posts (48.6%) in carousel format and 55 posts (51.4%) using unique images. On the other hand, posts in the events and internationalization categories stood out for their greater use of carousels, with 56.5 and 68.7% of posts, respectively, reflecting the need to present more than one element in a unique post.

The analysis of organization by category revealed that 70.7% of posts use unique images (278 posts), while 29.3% use carousels (115 posts), as we can see in the pie chart (Fig. 6).

6 Conclusions from the Results

The quantitative analysis allows us to identify relevant patterns in the activity of the universities, from which we highlight the following:

The volume of posts varied significantly between the universities. However, more posts do not necessarily result in more interactions, as evidenced by Instituto Superior Miguel Torga (ISMT), which had a high volume of posts but a low number of interactions per post, in contrast to Instituto Superior Manuel Teixeira Gomes, which, with the fewest posts, obtained the highest number of interactions per post.

Regarding the category of posts, it is noted that promotional posts were the most frequent, but they also had the lowest number of interactions.

Informational and recognition content proved more effective at generating interactions, suggesting that users value informative or celebratory content more. This also highlights the need for a shift in content strategy, as users are rejecting what they identify as promotional content.

In terms of visual characteristics performance, photographs generate more interactions than videos or combined formats, with textual content generating the fewest interactions.

Posts featuring original visual content received more interactions than those using stock image content; this underscores the importance of authenticity and content created specifically to convey a targeted message within the institutional context, making the audience feel the value of originality.

7 Conclusions

The conclusions of this study emphasize the importance of quality over quantity in Instagram content strategy. While frequent posting can increase visibility, it does not necessarily translate to higher engagement. Our study demonstrated that the university with the most posts had the fewest interactions. This finding reinforces the conclusions of Moreno-Albarracín and Blanco-Sánchez [7], who emphasize the importance of focusing on the quality of posts and consider that excessive posting tends to generate an adverse reaction from users, especially when the content lacks rigor and visual quality. The failure to adapt to what is expected in terms of Instagram communication leads to audience fatigue, resulting in disengagement and damaging the institution's image.

Thus, universities should produce authentic, high-quality content that resonates with their audience. Posts that humanize the institution by highlighting real stories and achievements foster a stronger sense of community and are more likely to generate interaction.

The importance of authenticity is reflected, for example, in ISMT, which achieved the highest interaction rates by posting photographs related to Internationalization (2 posts) and Participation in Events (4 posts). Similarly, Universidade Portucalense (UP) stood out by involving its students in the Global Management Challenge. Universidade da Maia (UM) reached 1,034 likes on a post about the Queima das Fitas parade. This last reference highlights the importance of exploring each university's cultural factors and unique features. Presenting institutional history and memories in a visual narrative adapted to the users and involving them whenever possible will be an effective strategy to promote the institution.

For private universities, Instagram offers the potential to build an engaged online community, but only if used effectively. Institutions should avoid overusing promotional content and instead explore storytelling approaches that highlight the human side of the institution. Visual elements, particularly original photography and simple, clean designs, are essential to capturing audience attention and driving interactions.

Regarding visual content, the study reveals that universities such as Europeia, Católica, Lusófona de Lisboa, and Portucalense create content specifically for the platform and present communication strategies consistent with a visual identity and positioning they seek to maintain.

Photographs are the type of post that generates the most interaction, while videos and combined formats generate few interactions. Authenticity and identification with the university remain vital factors in obtaining interactions, with stock image content consistently receiving fewer interactions.

Thus, it is concluded that original visual resources, explicitly created within the institutional context to convey specific messages, generate more interactions than any other resource type. The design should be striking but simultaneously simple and clean to allow for quick comprehension of the content. Text-based content generates the fewest interactions, reflecting the platform's purpose.

Creating an academic community on this platform can be seen as an "online campus." As such, it should promote interaction and engagement among the institution's members.

Further research is needed to explore the impact of Instagram Stories, Reels, and other emerging social trends. It could also examine user perspectives through interviews or sentiment analysis, as well as the role of hashtags and post timing in audience reach and engagement. Nevertheless, this study provides a solid foundation for future research and for understanding how private universities can optimize their social media strategies to create meaningful connections with their target audience.

Future Recommendations

Universities should:

- Humanize content by highlighting personal stories and events;
- Prioritize fewer but higher-quality posts focused on storytelling.
- Use original photography and clean visual designs to increase engagement.
- Balance promotional content with posts that showcase the university's culture, achievements, and student life.
- Encourage student and community engagement through participatory content.

References

1. DGES—Direção Geral do Ensino Superior (n.d.) Pesquisa de Cursos e Instituições. https://www.dges.gov.pt/Guias/indest.asp?reg=22. Last accessed 30 August 2024
2. OberCom (2024) Digital News Report: Portugal 2024. Reuters Institute for the Study of Journalism. https://obercom.pt/wp-content/uploads/2024/06/DNRP-2024-_-15.06.2024.pdf. Last accessed 15 July 2024
3. Kemp S (2024) Digital 2024: Portugal. https://datareportal.com/reports/digital-2024-portugal. Last accessed 15 July 2024
4. OberCom (2023) Retrato digital de Portugal: Caracterização e tendências de utilização das redes sociais 2015 a 2023. https://obercom.pt/retrato-digital-de-portugal-caracterizacao-e-tendencias-de-utilizacao-das-redes-sociais-2015-a-2023/. Last accessed 22 March 2024
5. Piza M (2012) O fenómeno Instagram: Considerações sob a perspetiva tecnológica. Instituto de Ciências Sociais, Departamento de Sociologia, Universidade de Brasília. https://bdm.unb.br/bitstream/10483/3243/1/2012_MarianaVassalloPiza.pdf. Last accessed 17 March 2024
6. Chester A (2018) The outmoded instant: from instagram to polaroid. Afterimage 45(5):10–15. https://doi.org/10.1525/aft.2018.45.5.10
7. Moreno-Albarracín B, Blanco-Sánchez T (2022) Comunidades de marca en Instagram: Estudio de las estrategias de las mejores universidades del mundo. Palabra Clave 25(4):e2545. https://doi.org/10.5294/pacla.2022.25.4.5
8. Alcolea Parra M, Rodríguez Barba D, Núñez Fernández V (2020) El uso corporativo de Instagram en las universidades privadas españolas. Estudio comparativo de treinta y cinco universidades. Ámbitos. Revista Internacional de Comunicación 47:109–134. https://doi.org/10.12795/Ambitos.2020.i47.06
9. Portaria n.º 256/2005, de 16 de março. https://diariodarepublica.pt/dr/detalhe/portaria/256-2005-572672. Last accessed 14 Aug 2024
10. Katz E, Blumler JG, Gurevitch M (1974) Utilization of mass media communication by the individual. In: Blumler JG, Katz E (eds) The uses of mass communications: current perspectives on gratifications research, vol 37. Sage, Beverly Hills, Califórnia, pp 509–523
11. Krippendorff K (1980) Content analysis: an introduction to its methodology. Sage, Newbury Park, CA